OBJECTIVE POULTRY SCIENCE

(for JRF, SRF, ARS, NET, SLET, Civil Services & Other Competitive Examinations)

THE AUTHOR

Dr. Santosh Haunshi *M.V.Sc., Ph.D.* Senior Scientist, ICAR, has completed the graduation (B.V.Sc & A. H.) from the Veterinary college, Bidar, Karnataka. He did his M.V.Sc., and Ph.D. from Indian Veterinary Research Institute, Izatnagar. He has been serving as scientist in ICAR for more than seven years. He is specialized in the area of poultry genetics and breeding. He is presently working as senior scientist at Project Directorate on Poultry, Rajendrangar, Hyderabad.

OBJECTIVE POULTRY SCIENCE

(for JRF, SRF, ARS, NET, SLET, Civil Services & Other Competitive Examinations)

Dr. Santosh Haunshi, *M.V.Sc., Ph.D.*

Senior Scientist

Project Directorate on Poultry

Rajendrangar, Hyderabad – 500 030

Daya Publishing House®

A Division of

Astral International Pvt. Ltd.

New Delhi – 110 002

ISBN 978-81-7035-714-8 (PB)

Published by : **Daya Publishing House**®
A Division of
Astral International Pvt. Ltd.
– ISO 9001:2015 Certified Company –
4736/23, Ansari Road, Darya Ganj
New Delhi-110 002
Ph. 011-43549197, 23278134
E-mail: info@astralint.com
Website: www.astralint.com

Digitally Printed at : **Replika Press Pvt. Ltd.**

Preface

Indian poultry industry is one of the vibrant and fastest growing agriculture sectors. India is number third in egg production and fifth in broiler production and the poultry industry directly or indirectly provides employment opportunity to about 1.5 million people in the country. By 2020 India could become closure to number one in poultry production in the world if it continues to grow at the present level and the necessary impetus is provided to the industry through adequate inputs including technical man power supply. Therefore, the requirement of technically qualified man power for the industry is enormous.

This book is intended for graduates opting for the poultry science or animal breeding discipline for their post graduation and for Ph.D. programs in the ICAR system, National eligibility Tests, Junior Research Fellowship, Senior Research Fellowship and Agricultural Research Service (ARS) examinations. Many of the competitive examinations mostly involve objective type questions. Therefore, keeping the requirements of students appearing for such competitive examinations in view, author has compiled and edited the objective type question bank in poultry science discipline.

Author hopes that the book will be useful not only to the students appearing for the almost all the competitive examinations in poultry science but also for teachers related to poultry science and animal breeding subjects.

Santosh Haunshi

Contents

Chapter 1

Poultry Production: An Overview

I. A. Multiple Choice Questions (Tick Correct Option/s)

Q. 1. Young ones of some birds have full coat of down feathers and are able to scurry about and find feed and water immediately after hatching, whereas young ones of some species of birds are helpless at the time of hatching and require constant attention and feeding by their parents. What is the technical word for the second groups of birds?

(*a*) Precocial birds (*b*) Altricial birds

(*c*) Homothermic birds (*d*) Noeteynal birds

Q. 2. Before receiving the chicks temperature under a brooder should be measured at what height?

(*a*) 6.5 cm above the litter (*b*) 4.5 cm above the litter

(*c*) 3.5 cm above the litter (*d*) 2.5 cm above the litter

Q. 3. What is high lethal body temperature for chicks?

(*a*) 105 °F (*b*) 117 °F

(*c*) 125 °F (*d*) 130 °F

Q. 4. What is the lower lethal temperature (internal body temperature) below which the day old chicks can not survive?

(*a*) 62 °F (*b*) 72 °F

(*c*) 82 °F (*d*) 92 °F

Q. 5. Fresh poultry droppings have how much moisture?

(*a*) 55-60% (*b*) 65-70%

(*c*) 75-80% (*d*) 80-85%

Q. 6. In cold season / climate how much cubic feet per minute of air flow per chick is sufficient for keeping the area dry under an electric brooder.

(*a*) 0.04 cfm (*b*) 0.10 cfm

(*c*) 0.30 cfm (*d*) 0.50 cfm

Q. 7. In a broiler house even after removing brooders what should be the minimum temperature inside the house

(*a*) 55-65°F (*b*) 70-75°F

(*c*) 80-85°F (*d*) 85-90°F

Q. 8. In a very large commercial broiler shed which system of brooding will be more convenient?

(*a*) Electric hovers

(*b*) Gas heated types

(*c*) Oil heated types

(*d*) Central heating system using hot water or hot air.

Q. 9. In the first week of brooding, the chick guard should be kept how much away from the edge of the hover.

(*a*) 6″-10″ (*b*) 12″-14″

(*c*) 18″-24″ (*d*) 24″-36″

10. When pullets are to be floor housed during the growing period, low roosts are installed at the age of 4 weeks. How much roosting space should be provided per chick at that age?

(*a*) 2″-3″ (*b*) 4″-5″

(*c*) 6″-8″ (*d*) 8″-10″

Q. 11. Light breeds pullets at maturity should get a minimum floor space of

(*a*) ½ sq. ft.–1sq.ft (*b*) 1 ¾ sq. ft.– 2 sq. ft

(*c*) 2–3 sq. ft. (*d*) 3–4 sq. ft

Q. 12. During rearing period for the pullets of heavy breeds restricted feeding is essential. The restricted feeding should start at what age?

(*a*) 4-6 weeks (*b*) 6-8 weeks

(*c*) 9-12 weeks (*d*) 13-16 weeks of age

Q. 13. Restricted feeding either provided free choice feeding on only alternate day or only a fraction of free choice feed, is given daily. The fraction to be cut in restricted feeding is

(*a*) Less than 5% (*b*) 5-10%

(*c*) 15-25% (*d*) 25-30%

Q. 14. Such restricted feeding is expected to restrict the growth of young pullets by

(*a*) 5-10% (*b*) 10-15%

(*c*) 15-20% (*d*) 20-25%

Q. 15. To avoid precocious sexual maturity in increasing natural day length, the light should be reduced up to 22 weeks of age at every 2 weeks by

(*a*) 20–30 minutes (*b*) 40–60 minutes

(*c*) 1 ½–2 hours (*d*) 2 ½–3 hours

Q. 16. After attaining sexual maturity layer should get a day length of

(*a*) 10-11 hrs (*b*) 11-12 hrs

(*c*) 14-16 hrs (*d*) 18 hrs

Q. 17. At least how much space under a brooder during first 3 weeks of age should be given to broilers?

(*a*) 7 sq. inches (*b*) 12 sq. inches

(*c*) 16 sq. inches (*d*) 18 sq. inches

Q. 18. How much minimum feeding space should be provided to broilers in tube feeders after 3 weeks of age?

(*a*) 1″ (*b*) 2″

(*c*) 3″ (*d*) 4″

Q. 19. How much body weight could be attained by Broad breasted Bronze toms at 24 weeks of age?

(*a*) 10 lbs (*b*) 12 lbs

(*c*) 24 lbs (*d*) 35 lbs

Q. 20. Broad breasted Hens can attain body weight at 17 weeks of age

(*a*) 10.5 lbs (*b*) 17.5 lbs

(*c*) 24.5 lbs (*d*) 30.5 lbs

I. B. Multiple Choice Questions (Tick Correct Option/s)

Q. 1. Which one of the following four strains belongs to broiler type?

(*a*) BV-300 (*b*) BV-380

(*c*) Babcock (*d*) ILI-80

Q. 2. Ovulation normally occurs:

(*a*) 30 minutes after laying

(*b*) 60 minutes after laying

(*c*) 2 hours after laying

(*d*) 24 hours after laying

Q. 3. The temperature of broody hen is relatively:

(*a*) Higher (*b*) Lower

(*c*) Sub-normal (*d*) No change

Q. 4. For minimum loss of feed, the feeder should be:

(*a*) Fully filled (*b*) 2/3rd filled

(*c*) ½ filled (*d*) 1/3rd filled

Q. 5. The body temperature of newly hatched chick is about:

(*a*) 38.7°C (*b*) 37. 7°C

(*c*) 36.7°C (*d*) 39.7°C

Q. 6. At an average humidity, birds start panting when the ambient temperature reaches:

(*a*) 85 °F (*b*) 75 °F

(*c*) 95 °F (*d*) 65 °F

Q. 7. For calculating the depreciation on poultry houses, the standard life span of the building should be:

(*a*) 20 years (*b*) 10 years

(*c*) 8 years (*d*) 5 years

Q. 8. For layers, the standard depletion of the stock up to the point of laying should be:

(*a*) 10% (*b*) 5%

(*c*) 3% (*d*) 1%

Q. 9. In adult chickens the body temperature varies between:

(*a*) 105 to 107 °F (*b*) 101 to 104 °F

(*c*) 98 to 101 °F (*d*) None of the above

Q. 10. The vegetable and animal protein in a poultry mash should be in the ratio of:

(*a*) 3:1 (*b*) 2:1

(*c*) 1:1 (*d*) 1:3

Q. 11. The upper lethal body temperature of adult chicken is about:

(*a*) 47 °C (*b*) 46 °C

(*c*) 45 °C (*d*) 43 °C

Q. 12. The poultry feed becomes susceptible to moulds and becomes overheated if its moisture content is more than:

(*a*) 8% (*b*) 10%

(*c*) 12% (*d*) 13%

Q. 13. Which one of the following feed ingredients is highly potent source of afflatoxin?

(*a*) Rice polish
(*b*) Fish meal
(*c*) Ground nut cake
(*d*) Wheat bran

Q. 14. The vitelline membrane which holds the fluid yolk mass together is:

(*a*) Red (*b*) Yellowish
(*c*) Pale green (*d*) Colourless

Q. 15. What is the ideal ambient temperature for laying hens?

(*a*) 32°F to 50°F (*b*) 55°F to 75°F
(*c*) 80°F to 90°F (*d*) 95°F to 105°F

Q. 16. How much manure will be produced by 100,000 layers per day?

(*a*) 3 tons (*b*) 6 tons
(*c*) 9 tons (*d*) 12 tons

Q. 17. Which of the following species of birds is a seasonal layer?

(*a*) Chicken (*b*) Ducks
(*c*) Quails (*d*) Guinea fowls

Q. 18. The eggs of Mediterranean class of chicken are:

(*a*) White (*b*) Blue
(*c*) Yellow (*d*) Brown

Q. 19. The biological value of egg protein is:

(*a*) 95 (*b*) 90
(*c*) 85 (*d*) 80

Q. 20. Which one of the following body parts shed the feathers first during molting?

(*a*) Head (*b*) Neck
(*c*) Breast (*d*) Wings

Q. 21. Which of the following body parts is depigmented first when the pullets starts laying?

(*a*) Eye ring (*b*) Vent

(*c*) Earlobes (*d*) Shank

Q. 22. The maximum width of a brooder house should be:

(*a*) 30 ft. (*b*) 50 ft.

(*c*) 70 ft. (*d*) 90 ft.

II. Fill in the Blanks

1. India ranks_______in broiler production in the world
2. India ranks_______in egg production in the world
3. The present annual egg production of India is_______billions
4. The annual broiler production of India is_______millions
5. The per capita availability of egg in India is_______
6. _______is the albumen secreting portion of the oviduct.
7. Any cessation of laying for seven or more days during winter months is called_______
8. _______are small chicken, usually weighing from 0.45-0.85 kg, which are generally used for show and as pets.
9. The class of poultry with yellow beaks, skin, shanks and toes is known as_______.
10. The distinctive character of the mouth of bird is the absence of_______and_______.
11. The grinding and crushing process of the gizzard in fowl is aided by the presence of_______or_______taken through the mouth.
12. The newly hatched chicks should be placed on water_______hrs prior to free choice feed.

13. The ideal moisture content of the litter of a poultry shed should be_______to_______%.

14. Though the hens are capable to produce eggs at higher and lower temperatures but moderate temperature is_______°C with adequate diet.

15. There are two blind pouches at the junction of lower small intestine and rectum called_______.

III. Write True or False for the Following Statement

1. Building large poultry houses increases cost of construction

2. The temperature of the Poultry house should be 95 °F during the first two weeks of brooding period

3. The daily water requirement of 100 laying hens is 3 gallons

4. Abrupt change in feed in laying hens results in fall in consumption of feed and corresponding loss in production.

5. Pullets should be fed a diet containing 15 to 16 per cent protein.

6. Trap nest is essential equipment for pedigree breeding, where one male is continuously mated to one female in a breeding pen.

7. Individual laying cages are mainly used to solve the cannibalism in layers.

8. The purpose of caponizing is to improve the quality of poultry meat.

9. The desirable pH for ideal litter in poultry is 8.0 or more.

10. For preserving the fertilizer value of poultry manure Di-Ammonium Phosphate should be added.

11. The molting order in laying hen is head, neck, breast, back, fluff, abdomen, wings and tail.

12. Conversion of protein with lower biological value to that in higher biological value does not take place in poultry.

13. The digestive system of poultry is adapted to the utilization of large quantities of fibrous food.

14. A certain amount of indigestible organic matter is necessary to maintain the tone of the gut.

15. The energy protein ratio as per ISI (1992) standard for starter (broiler) should be 2800:23.

16. The lysine and methionine are critical amino acids for poultry.

17. The ratio of vitamin content to digestible protein is as much critical as that of the amino acids to digestible protein in the diet of poultry.

18. The deficiency of calcium in the diet of poultry affects the absorption and utilization of phosphorus in the poultry.

19. Pelleting of poultry feed is disadvantageous over mash feed.

20. The egg shell of poultry is loosely covered with chalky powder known as cuticle which physically blocks the entrance of microorganisms.

21. The chalaziferous layer of thick albumin surrounds the yolk.

22. Washed and dry cleaned eggs preserve will

Answers

I. A. Answer for Multiple Choice Questions

1	(*b*)	6	(*b*)	11	(*b*)	16	(*c*)
2	(*b*)	7	(*b*)	12	(*c*)	17	(*a*)
3	(*b*)	8	(*d*)	13	(*c*)	18	(*c*)
4	(*a*)	9	(*d*)	14	(*b*)	19	(*d*)
5	(*c*)	10	(*b*)	15	(*b*)	20	(*b*)

I. B. Answer for Multiple Choice Questions

1	(*c*)	7	(*a*)	13	(*c*)	19	(*a*)
2	(*a*)	8	(*b*)	14	(*d*)	20	(*a*)
3	(*a*)	9	(*a*)	15	(*b*)	21	(*a*)
4	(*b*)	10	(*a*)	16	(*d*)	22	(*a*)
5	(*d*)	11	(*a*)	17	(*d*)		
6	(*a*)	12	(*c*)	18	(*a*)		

II. Answers for Fill in the Blanks

1. Fifth
2. Third
3. 50–55 billion
4. 400-600 million
5. 40-45 eggs per-capita availability
6. Magnum

7. Winter pause
8. Bantams
9. Mediterranean class
10. Tooth and taste buds
11. Grits/stones
12. 6 Hours
13. 20-25%
14. 24°C
15. Ceca

III. Answers for True or False

1. False
2. False
3. True
4. True
5. True
6. False
7. True
8. False
9. True
10. False
11. True
12. False
13. False
14. True
15. True
16. True
17. False
18. True
19. False
20. True
21. False
22. False

Chapter 2

Fundamentals of Poultry Genetics and Breeding

I. A. Multiple Choice Questions (Tick correct option/s)

Q. 1. Which jungle fowl has a single median wattle

(*a*) Red Jungle fowl (*b*) Ceylon jungle fowl

(*c*) Grey jungle fowl (*d*) Javan jungle fowl

Q. 2. White leghorn breed cannot be used as a silver female in silver x gold cross because it is having

(*a*) White ear lobes (*b*) Single comb

(*c*) Dominant white (*d*) Yellow beak

Q. 3. Rumplessness is due to

(*a*) Autosomal recessive

(*b*) Autosomal dominant

(*c*) Sex-linked dominant

(*d*) Sex-linked recessive

Q. 4. Silkiness is due to

(*a*) Autosomal recessive

(*b*) Autosomal dominant

(*c*) Sex-linked dominant

(*d*) Sex-linked recessive

Q. 5. Frizzling is due to

(*a*) Autosomal recessive

(*b*) Autosomal dominant

(*c*) Sex-linked dominant

(*d*) Sex-linked recessive

Q. 6. When blue fowls are mated together they produce black, blue and blue splashed whiter birds in the ratio of

(*a*) 3:1:0 (*b*) 1:2:1

(*c*) 2:1:1 (*d*) 1:1:2

Q. 7. Birds of white leghorn breed are white because they have

(*a*) No colour gene

(*b*) A dominant gene that inhibits colour

(*c*) They have recessive white gene

(*d*) They have silver gene

Q. 8. The best breed for using as male line in broiler production is

(*a*) White Rock (*b*) Cornish

(*c*) New Hampshire (*d*) Australorp

Q. 9. The White and Black barring in Silver Campine is due to

(*a*) Autosomal barring gene

(*b*) Sex-linked barring gene

(*c*) Both A and B

(*d*) None of the above

Q. 10. The Black and Gold barring in the Gold Campine is due to

(*a*) Autosomal barring gene

(*b*) Sex-linked barring gene

(*c*) Both (*a*) and (*b*)

(*d*) None of the above

Q. 11. The Black and White barring in Barred Plymouth Rock is due to

(*a*) Autosomal barring gene

(*b*) Sex-linked barring gene

(*c*) Both (*a*) and (*b*)

(*d*) None of the above

Q. 12. The chromosome number of chicken is

(*a*) 78 (*b*) 80

(*c*) 82 (*d*) 86

Q. 13. Dual purpose white Plymouth Rock belongs to

(*a*) Asiatic class (*b*) American class

(*c*) English class (*d*) Mediterranean class

Q. 14. White Ear-lobe is one of the most important characteristics of

(*a*) Asiatic class (*b*) American class

(*c*) English class (*d*) Mediterranean class

Q. 15. *Gallus gallus* is popularly known as

(*a*) Ceylon jungle fowl (*b*) Grey jungle fowl

(*c*) Javan jungle fowl (*d*) Red jungle fowl

Q. 16. Commercial hybrid layers are usually the product of

(*a*) Line crosses (*b*) Strain crosses

(*c*) Inbred hybrids (*d*) All the above

Q. 17. Which of the following breeds are commonly used as a female line for production of commercial broilers?

(*a*) New Hampshire (*b*) Ply mouth Rock

(*c*) Both (*a*) & (*b*) (*d*) None of the above

Q. 18. Darwin the Naturalist is famous for the theory of

(*a*) Inheritance of acquired characters

(*b*) Formation of species

(*c*) Natural selection

(*d*) Quantitative genetic theory

Q. 19. The scientific name of red jungle fowl is

(*a*) *Gallus gallus*

(*b*) *Gallus lafayetti*

(*c*) *Gallus sonnerati*

(*d*) *Gallus varius*

II. Fill in the Blanks

1. The study of birds which are not classified as poultry is called_______
2. The term_______is used to designate group of breeds which have been developed in certain regions or geographical areas
3. Zoologically the fowl belongs to_______Kingdom
4. The chicken belongs to_______Phylum
5. The chicken belongs to_______Class
6. The chicken belongs to the_______Genus
7. The chicken belongs to_______Family
8. Birds of distinct type and colour patterns admitted to the standard are termed as_______
9. With 2 pairs of genes the F2 ratio is_______
10. With 3 pairs of genes the F2 ratio is_______
11. The non-allelic gene interaction is called_______

12. The White Plymouth Rock is not barred although it has the barring gene because the barring can be expressed only when there is a_______gene.
13. The gene for barring inhibits deposition of_______
14. The sex-linked dwarf gene is incompletely_______to normal body size.
15. The skin colour in the domestic fowl arises from the presence of the_______and_______Pigments.
16. In Poultry male sex is_______
17. In Poultry female sex is_______
18. The best breed for using as male line in broiler production is_______
19. Vanaraja is a dual purpose variety of fowl developed for _______system of rearing
20. _______is the ancestor of modern day fowl.
21. Inbreeding increases_______in a population.
22. _______generally reduces performance, since undesirable genes are concentrated in the inbred lines.
23. When a particular strain in a cross nearly always gives a good result, the strain is said to have_______
24. The ability of quantitative characters to be transmitted from parent to offspring is known as_______
25. The Javan jungle fowl has an extra pair of_______
26. The Brahma breed of chicken has a_______comb
27. Some breeds are divided into varieties according to colour/ type of_______
28. The colour of earlobes is_______in Mediterranean breeds.
29. All chicken breeds belonging to English class lay brown shelled eggs except_______breed.

30. Leghorn breed of fowl is the world's number one egg producer that originated from_______
31. _______is the mating between birds, which are more closely related to each other than average of a population.
32. Mating between sibs or parent and progeny to form inbred lines is called_______
33. Inbreeding within an ancestral line with the object to increase the particular male or female ancestor's proportions in the progeny is called_______
34. Rate of feathering is an autosomal character and slow rate of feathering is_______over fast rate.
35. _______are the progeny of two inbred lines of different breeds.
36. Plymouth Rock is a chosen American breed because of its egg size and_______properties.
37. White Cornish is mostly used as_______line parent for better broiler production.
38. _______comb is the breed characteristic of Brahma chicken.
39. The Javan jungle fowl is Gallus_______
40. The Austrolap breed belongs to_______class.
41. The Jersey Black Giant breeds belongs to_______class.
42. The Orpington breed belongs to_______class.
43. The Langshan breed belongs to_______class.
44. The breed that lays blue shelled eggs is_______
45. The hen feathered breed is_______
46. An established group of birds having the same general body shape, weight and some common characteristics is called_______

47. A sub-division of a breed, distinguished either by colour pattern, shape, comb type or feather pattern is called______

48. The known species of wild fowl belong to the genus called______

49. The grey jungle fowl is G______

50. The Javan jungle fowl is G______

51. The Ceylon jungle fowl is G______

52. Mention the two theories of origin of fowl

    (*a*) ______ (*b*) ______

53. The Jersey Black Giant breed belongs to______class.

54. The Orpington breed belongs to______class.

55. The Australorp breed belongs to______class.

56. All English class breeds lay brown shelled eggs except______and______breeds.

57. All American class breeds lay brown eggs except______which lay White eggs.

58. The Rhode Island Red breed has______ and______ comb varieties.

59. List the names of the seven varieties of Plymouth Rock.

60. The *Langshan* breed belong to______class.

61. The *Hondan* breed has______toes.

62. The hen-feathered breed is______

63. Name the four varieties of *Aseel* breed.

64. The nonlinear interaction between any non-allelic genes is called______

65. The rose comb is due to a single gene but the points on the comb are determined by the effect of_______genes.

66. Genes that prevent the phenotypic effect of other genes are called_______genes.

67. Characters expressed only by one sex are called_______

68. Characters transmitted from dam to son are called_______

69. The barring in Silver and Golden Campine is due to_______gene.

70. The gene for barring_______deposition of_______and therefore cause White bars on black feathers.

71. The White Leghorns are silver but not useful in silver and gold cross because of_______gene.

72. The modifiers of early feathering are_______and exert its effect on_______flight feathers.

73. Birds with modifier (t gene) of early feathering have normally developed_______feathers but no_______ feathers at hatching time.

74. The thyrogenous dwarfism is due to a gene that is_______and_______

75. Some breeds have yellow shanks as they lack_______in the dermis because they carry_______gene.

76. The genotype of coloured shanked breeds is_______

77. An example of chicken breed with dominant white gene is_______

78. The common name of *Gallus lafayettii* is_______and is characterized by_______secondary feathers in females.

79. *Gallus varius* is referred to as_______Jungle Fowl and has a_______wattle.

80. The breeds commonly used for auto sexing are_______and_______

81. The pioneer worker, who reported_______rumplessness, was_______in the year_______referring to_______ condition.

82. The four known species of wild fowls belong to the genus called_______

83. Name four species of wild fowls

(*a*) _______	(*b*) _______
(*c*) _______	(*d*) _______

84. Leghorn and other Mediterranean breeds closely resemble *Gallus*_______

85. Retarded and tardy are the autosomal modifiers of_______

86. In a sex-linked cross involving barring, the female parent is a_______

87. In a sex-linked cross involving silver and gold the silver gene carrying_______parent is used.

88. In a sex-linked cross involving silver and gold, the gold gene carrying_______parent is used.

89. In a sex-linked cross involving barring the chicks with a white spot on head and yellow beak, shanks and toes belong to_______sex.

90. In a sex-linked cross involving feathering gene early feathering_______parent is used.

91. In a sex-linked cross involving silver and gold, the chicks with buff or red colour belong to_______sex.

92. In a sex-linked cross involving feathering gene, the chicks with at least seven visible secondaries are separated as_______chicks.

III. A. Write True or False for the Following Statement

1. Leghorn and other Mediterranean breeds closely resemble *Gallus gallus*.
2. The Asiatic class chickens always lay brown eggs.
3. The Cochin breed of chicken has feathered shanks.
4. The autosomal lethal gene eliminates the heterozygous individuals.
5. The Australorp male and White Leghorn female cross-shown excessive broodiness.
6. Project Directorate on Poultry is the Coordinating centre of AICRPs on Poultry.
7. At the age of ten days slow feathering males and females do not have any visible tail.
8. The width of bars on plumage in sex-linked barring is same in both the sexes.
9. White Plymouth Rock breed is responsible for origin of modern day commercial layer strains.
10. The mating of early feathering males with early feathering females will produce only slow feathering females.
11. Modern domestic breeds are considered to be the descendents of *Gallus lafayetti*.
12. Javan Jungle fowl is known as *Gallus varius*.
13. *Gallus sonneratii* is known as Red Jungle fowl.
14. Brahma, Cochin and Langshan are the 3 recognized breeds of Asiatic Class.
15. *Gallus gallus* is popularly called as Ceylon jungle fowl.
16. The chicken breeds of American class are characterized by yellow skin, clean shanks, and read ear lobes.

17. White Plymouth Rock breed is responsible for origin of modern day commercial layer strains.
18. Central Avian Research Institute is a coordinating centre of AICRP on Poultry.
19. Egg production is a sex- limited character.
20. Diploid Chromosome number of chicken is 76.
21. White Leghorn is white because it does not carry the coloured genes.
22. Slow feathering gene in chickens is recessive to rapid feathering.
23. The genetic code is triplet code.
24. Most of the commercial broilers and layers sold in the market are crosses of one kind or the other.
25. The Asiatic breeds of chicken have red ear lobes.
26. The eggs of Ceylon jungle fowl are spotted.
27. The White Pekin is egg type breeds of ducks.
28. The eggs of Ceylon jungle fowl are spotted.
29. The grey jungle fowl has a smooth edged comb.
30. The Javan jungle fowl has a median wattle.
31. The American class breeds lay only white eggs.
32. The Cornish breed belongs to American class.
33. The American class have feathered shanks.
34. The Asiatic class lay brown eggs.
35. The Asiatic breeds have red ear lobes.
36. The Cochin breed has feathered shanks.
37. The RIR male and White Leghorn female cross mature early and do not show broodiness.

38. The covert is about 2/3 as long as and more slender than primary feather in slow feathering phenotype at the time of hatch.
39. No visible secondary feathers are seen in fast feathering phenotype at the time of hatching.
40. So far 13 blood group systems have been identified in chickens.
41. The 'B' blood group system serves as a marker for major histo-compatibility complex.
42. In random mating populations, the zygotic ratio is the square of the gametic ratio.
43. Test crossing and back crossing are same.
44. The dwarfing gene reduces the female body size by about 15 percent.
45. *Gallus soneratii* has genes for sex-linked silver.
46. Barring in poultry is only sex-linked.
47. Barred males (sex-linked) show whiter plumage in the homozygous than hemizygous females (sex linked).
48. Rose comb is dominant to pea comb.
49. White Leghorns are barred.
50. Spangled feather pattern is reverse of mottling feather pattern.
51. Pied feather pattern is a breed characteristic of Exchequer Leghorns.
52. The function of gene for barring is restriction of melanin pigmentation.
53. Gene for dominant white (homozygous) in Canaries is non lethal.
54. Anadidymous is a non-genetic character.

III. B. State True or False for the Following Statements

1. Pea comb did occur in some of the wild species of jungle fowl.
2. The characteristics of Asiatic breeds are found in all species of jungle fowl.
3. Khaki Campbell ducks need water for swimming.
4. The width and sharpness of barring is same in the two sexes.
5. For Columbian pattern the sex linked gene silver and an autosomal gene are needed.
6. Dwarf chicks can be separated from normal chicks only at day old at the time of hatching.
7. Early feathering and slow feathering chicks can be separated also at 10 days of age.
8. Thyrogenus dwarfism is due to recessive autosomal dwarf gene.
9. It is difficult to classify young birds as white skinned or yellow skinned, until they are three months of age.
10. The sex-linked recessive gene responsible for melanin in the dermis is useful for identification of sexes at hatching in sex-linked crosses.
11. The White Leghorns also possess genes for colour and barring.
12. The Cornish breed has White skin.
13. American breeds have red ear lobes.
14. Malay breed has feathered shanks.
15. Blue plumage is due to heterozygous condition of colour genes.

16. Rose comb is dominant to Walnut comb.

17. Pea comb is dominant to Walnut comb.

18. Naked neck is incompletely dominant to normal neck feathering.

19. Feathered shanks are recessive to non-feathered shanks.

20. Creeper condition in chickens is dominant to normal condition.

21. Walnut comb is dominant to rose, pea and single comb.

22. Normal tail is incompletely dominant to long tail.

23. The first auto sexing breed emerged from a curious attempt of mating of sex-linked and autosomal barring gene carrying parents.

24. The flocks supplied to hatchery man are referred as parent lines.

25. The foundation breeder is only concerned with the genetic improvement of pure lines.

26. In a three way cross the female parent is a single cross.

27. In a three way cross the male parent is a single cross.

28. In a four way cross the female parent is a pure line.

29. In a four-way cross the male parent is also a single cross.

30. The hatchery man in a franchise arrangement is supplied only males of male line and females of female line.

31. Breed crosses for broiler production involve strains representing different breeds.

32. Inbred hybrid are the progeny produced from one inbred line and one non-inbred line

IV. A. Match the Following

1.	American Breed	(*a*)	Aseel
2.	Asiatic Breed	(*b*)	Australorp
3.	Auto-sexing Breed	(*c*)	Aylesbury
4.	Desi Breed	(*d*)	Brahma
5.	English Breed	(*e*)	Campbell
6.	Egg type Duck Breed	(*f*)	Crested white
7.	Mediterranean Breed	(*g*)	Cambars
8.	Meat type Duck Breed	(*h*)	White Holland
9.	Ornamental type Duck Breed	(*i*)	Minorca
10.	Turkey Variety	(*j*)	Vanaraja
		(*k*)	Wyandotte

IV. B. Match the Autosexing Breeds with their Developers

1.	Cambars	(*a*)	Jaap (1940)
2.	Oklabars	(*b*)	Punnet and Pease (1930)
3.	Legbars	(*c*)	Punnet (1940)
4.	Dorbars	(*d*)	Lamoreux (1941)
5.	Ancobars	(*e*)	Pease (1941)

Answers

I. Multiple Choice Questions

1	(*d*)	6	(*b*)	11	(*b*)	16	(*d*)
2	(*c*)	7	(*b*)	12	(*a*)	17	(*c*)
3	(*b*)	8	(*b*)	13	(*b*)	18	(*c*)
4	(*a*)	9	(*a*)	14	(*d*)	19	(*a*)
5	(*b*)	10	(*a*)	15	(*d*)		

II. Fill in the Blanks

1. Ornithology
2. Class
3. Animalia
4. Chordata
5. Aves
6. Gallus
7. Phasianidae
8. Standard Breed
9. 9:3:3:1
10. 27:9:9:9:3:3:3:1
11. Epistasis
12. Colour
13. Melanin
14. Recessive

15. Melanin and Xanthophyll
16. Homogametic
17. Heterogametic
18. Cornish
19. Backyard
20. Red jungle fowl
21. Homozygosity
22. Inbreeding
23. General Combining Ability
24. Heritability
25. Rectrices or tail feathers
26. Pea
27. Feather /Comb
28. White
29. Dorking/ Red cap
30. Italy
31. Inbreeding
32. Close Inbreeding
33. Line Breeding
34. Dominant
35. In-crossbreds
36. Fleshing/Meat
37. Male
38. Pea
39. varius
40. English
41. American
42. English
43. Asiatic
44. Araucauna

45. Campine
46. Breed
47. Variety
48. Gallus
49. *sonneratti*
50. various
51. *lafayetti*
52. (*a*) Monophyletic origin, (*b*) Polyphyletic origin
53. American
54. English
55. English
56. Dorking and Red Cap
57. Holland
58. Single, rose
59. (*a*) White, (*b*) Barred, (*c*) Buff, (*d*) Silver penciled, (*e*) Partridge, (*f*) Columbian, (*g*) Blue
60. Asiatic
61. Five
62. Campine
63. (*a*) Peela, (*b*) Kagar, (*c*) Chitta, (*d*) Nurie
64. Epistasis
65. Modifying
66. Inhibiting
67. Sex limited
68. Sex-linked
69. Autosomal barring
70. Inhibit, melanin
71. Dominant White
72. Autosomal, secondary
73. Primary, secondary

74. Recessive, autosomal
75. Melanin, Id
76. CCEE
77. White Cornish
78. Ceylonese and barred
79. Javan, median
80. White Leghorn and Rhode Island Red or New Hampshire
81. Dominant, Aldrovandus, 1600, tail-less-ness.
82. *Gallus*
83. (*a*) *Gallus gallus,* (*b*) *Gallus lafayetti,* (*c*) *Gallus sonneratti,* (*d*) *Gallus varius*
84. *gallus*
85. Sex-linked feathering gene
86. Barred Plymouth Rock
87. female
88. male
89. female
90. male
91. male
92. female

III. A. True or False

1.	True	9.	False	17.	False	25.	True
2.	False	10.	False	18.	False	26.	True
3.	True	11.	False	19.	True	27.	False
4.	False	12.	True	20.	False	28.	True
5.	False	13.	False	21.	False	29.	False
6.	True	14.	True	22.	False	30.	True
7.	False	15.	False	23.	True	31.	False
8.	False	16.	True	24.	True	32.	False

33.	False	39.	False	45.	False	51.	True
34.	True	40.	True	46.	False	52.	True
35.	True	41.	True	47.	True	53.	False
36.	True	42.	True	48.	False	54.	True
37.	False	43.	False	49.	True		
38.	False	44.	False	50.	True		

III. B. True or False

1	False	10	False	19	False	28	False
2	False	11	True	20	True	29	True
3	False	12	False	21	True	30	True
4	False	13	True	22	False	31	True
5	True	14	False	23	True	32	True
6	False	15	True	24	True		
7	True	16	False	25	True		
8	True	17	False	26	True		
9	True	18	True	27	False		

IV. A. Answers for Match the Following

1.	(*k*)	2.	(*d*)	3.	(*g*)	4.	(*a*)
5.	(*b*)	6.	(*e*)	7.	(*i*)	8.	(*c*)
9.	(*f*)	10.	(*h*)				

IV. B.

1.	Cambars	(*b*)	Punnet and Pease (1930)
2.	Oklabars	(*a*)	Jaap (1940)
3.	Legbars	(*c*)	Punnet (1940)
4.	Dorbars	(*e*)	Pease (1930)
5.	Ancobars	(*d*)	Lamoreux (1941)

Chapter 3

Inbreeding, Heritability and Correlations

I. A. Multiple Choice Questions (Tick Correct Answer from Choices)

Q. 1. The difference between the genotypic value (G) and the breeding value (A) of a particular genotype is known as

(*a*) Dominance deviation (*b*) No dominance

(*c*) Directional dominance (*d*) None of the above

Q. 2. A specific hypothesis is about a population that is being tested by means of the sample result is termed as

(*a*) Sampled population (*b*) Test of significance

(*c*) Frequency distribution (*d*) Null hypothesis

Q. 3. Heritability of character in chicken can be increased by

(*a*) Crossing the flock with new diversified genetic material

(*b*) Inbreeding the individuals

(*c*) Increasing environmental variation

(*d*) None of the above

Q. 4. The qualitative traits are governed by genes having large phenotypic effects and may be called

(*a*) Quantitative genes (*b*) Major genes

(*c*) Minor genes (*d*) Poly genes

Q. 5. The theoretical basis of quantitative genetics was established by the innovation and famous work of

(*a*) J. B. S. Haldane (*b*) Sewall Wright

(*c*) R. A. Fisher (*d*) All the above

Q. 6. An individual when mated to a number of individuals taken at randomly from the population, the mean deviation of the progeny from the population mean will be

(*a*) Equal to its breeding value

(*b*) Half of its breeding value

(*c*) Double to its breeding value

(*d*) None of the above

Q. 7. For the analysis of non-orthogonal data with unequal sub-class numbers a linear statistical model was given by

(*a*) Henderson (1949) (*b*) Kampthorne (1957)

(*c*) Harvey (1975) (*d*) All the above

Q. 8. In general, the consequences of genotype-environmental interactions appear to be more serious in

(*a*) Egg production (*b*) Broiler stock

(*c*) Growth (*d*) All the above

Q. 9. Depending on the distribution, environmental effects can be classified either a continuous or discrete. The discrete types are

(*a*) Micro-environment

(*b*) Macro-environment

(*c*) Intra-population variance

(*d*) Inter-population variance

Q. 10. Only the phenotypic values of individuals can be directly measured but the value that determines their influence of the next generation is

(*a*) Breeding value (*b*) Genotypic value

(*c*) Phenotypic value (*d*) Genome

Q. 11. The correlation between breeding values and phenotypic values is equal to the square root of the

(*a*) Standard error

(*b*) Coefficient of variation

(*c*) Sampling variance

(*d*) Heritability (Narrow sense)

Q. 12. The precision of the heritability estimates depends on its

(*a*) Phenotypic correlation (*b*) Genetic correlation

(*c*) Environmental correlation (*d*) Correlated characters

Q. 13. In principle, there are correlations between dominance deviations and between the various interaction deviations which is called

(*a*) Phenotypic correlation (*b*) Genetic correlation

(*c*) Environmental correlation (*d*) Correlated characters

Q. 14. In an analysis of half–sib families the component of covariance between sire is

(*a*) ½ Cov_A (*b*) ¼ Cov_A

(*c*) ¾ Cov_A (*d*) Cov_{AB}

Q. 15. Correlation between repeated measurements of the same individual is called

(*a*) Heritability (*b*) Realized Heritability

(*c*) Repeatability (*d*) Breeding value

Q. 16. Variation without natural discontinuities is called continuous variation and character that exhibit it are called

(*a*) Metric character (*b*) Correlated character

(*c*) Neutral character (*d*) Balanced character

17. The resemblance between offspring and parents provides the basis for

(*a*) Discrete breeding (*b*) Inbreeding

(*c*) Random breeding (*d*) Selective breeding

Q. 18. The precision of a heritability estimate is known by the

(*a*) Standard error (*b*) Magnitude

(*c*) Method of estimation (*d*) Exponential design

Q. 19. The genetic variance will be maximum when the gene frequency is

(*a*) 0.1 (*b*) 0.5

(*c*) 0.9 (*d*) 1.0

Q. 20. When there is tight linkage in the coupling phase the expected proportion of F_2 individuals of (Ab) type is

(*a*) 1/4 (*b*) 1/2

(*c*) 3/4 (*d*) 0

I. B. Multiple Choice Questions (Tick Correct Answer from Choices)

Q. 1. The consequences of regular system of inbreeding were first described in detail by

(*a*) Wright (1921) (*b*) Fisher (1949)

(*c*) Cruden (1949) (*d*) Haldane (1937)

Q. 2. The most rapid method of attaining high level of inbreeding in chickens

(*a*) Selfing (*b*) Full-sib mating

(*c*) Father-daughter mating (*d*) Half-sib mating

Q. 3. The number of generations of full brother sister mating required to produce progeny with 50 percent inbreeding coefficient

(*a*) 2 (*b*) 3

(*c*) 4 (*d*) 5

Q. 4. The character that is least affected by inbreeding depression

(*a*) Egg production (*b*) Hatchability

(*c*) Body Weight (*d*) Mortality rate

Q. 5. By saying that a character shows inbreeding depression, we refer to the

(*a*) Reduction in the fitness of the flock

(*b*) Fall in the productivity of some lines

(*c*) Average change of mean value over all the lines

(*d*) Loss of heterozygosity

Q. 6. Inbreeding depression is a consequence due to change in

(*a*) Genotypic frequency

(*b*) Gene frequency

(*c*) Genetic variability

(*d*) Phenotypic variability

Q. 7. General consequences of inbreeding in a population is

(*a*) Decrease of within line and between line variance

(*b*) Decrease of within line variance and increase of between line variance

(*c*) Increase of within line and between line variance

(*d*) No change of variance either within lines of between lines.

Q. 8. The number of years required to attain an inbreeding coefficient of 90 percent in chickens.

(*a*) 10 (*b*) 11

(*c*) 5 (*d*) 6

Q. 9. When the sexes are unequal in number in a poultry breeding flock the rate of inbreeding is given by the formula F= 3/32Nm +1/32Nf according to

(*a*) Wright (*b*) Crow

(*c*) Gowe, Robertson and Latter (*d*) Malecot

Q. 10. Reciprocal recurrent selection is method for improvement of

(*a*) S.C.A.

(*b*) G.C.A.

(*c*) S. C. A. and G.C.A.

(*d*) Egg production traits.

Q. 11. In a population that is in equilibrium the frequencies of the genotypes, D, H, R were (4, 32, 64), respectively. What will be the frequency of mating of AA × aa?

(*a*) 0.512 (*b*) 0.256

(*c*) 0.0512 (*d*) 0.0256

Q. 12. The percentage of heterozygotes in a large random mating flock of chickens showing 9 percent single combed birds and 91 percent rose combed birds.

(*a*) 42 (*b*) 0.198

(*c*) 0.03 (*d*) 0.0256

Q. 13. The method of computing inbreeding coefficient on the basis of co-ancestry was proposed by

(*a*) Cruden (*b*) Wright I. A.

(*c*) Cruden, Emik and Terrill (*d*) Lerner

Q. 14. The scientist not connected to diallel cross studies

(*a*) Hayman, B. I. (*b*) Jinks, J. L.

(*c*) Griffing B. (*d*) Hazel L. N.

Q. 15. The earliest workers who did diallel crossing in chickens

(*a*) Yao T. S. (*b*) Golo & Nordskog

(*c*) Wearden & Tindell (*d*) Hill and Nordskog

Q. 16. Diallel crossing studies in poultry indicate that the most important genetic source of variation for most of the traits.

(*a*) G. C. A. (*b*) S. C. A.

(*c*) Maternal effects (*d*) Sex linked effects

Q. 17. In a population, where in 'A' is dominant over 'a' and all the matings are between individuals of the same phenotype, that is, dominants mate only with dominants and recessives mate only with recessive (complete positive assortative mating) when we compare the parental population with the offspring population.

(*a*) There will be no change in gene frequency

(*b*) There will be no change in genotypic frequency

(*c*) There will be no change in phenotypic frequency

(*d*) There will be no change in all the above three

Q. 18. In a program of recurrent selection for specific combining ability what should be the nature of the tester population

(*a*) Synthetic population

(*b*) Random bred

(*c*) Inbred

(*d*) Similar to the selected line

Q. 19. Heterosis on crossing can result when there is

(*a*) Dominance (*b*) No Dominance

(*c*) Directional dominance (*d*) Inbreeding

Q. 20. Characters insensitive to inbreeding

(*a*) Reproductive traits (*b*) Maternal traits

(*c*) Skeletal traits (*d*) Fitness traits

I. C. Multiple Choice Questions (Tick Correct Answer from Choices)

Q. 1. The value observed, when a character is measured on an individual is called

(*a*) Genotypic value (*b*) Phenotypic value

(*c*) Breeding value (*d*) Average value

Q. 2. Parents pass on these to the next generation

(*a*) Genes (*b*) Genotypes

(*c*) Phenotypes (*d*) Gametes

Q. 3. The most common method followed in estimating heritability of traits in chickens

(*a*) Full-sib correlation

(*b*) Half-sib correlation

(*c*) Regression of offspring on parent

(*d*) Identical twin data analysis

Q. 4. The term heritability means

(*a*) Proportion of genetic variance

(*b*) Proportion of additive genetic variance to environmental variance

(*c*) Ratio of additive genetic variance to phenotypic variance

(*d*) Regression of phenotypic value on breeding value

Q. 5. Character of high heritability in chickens

(*a*) Age at first egg (*b*) Viability

(*c*) Annual egg production (*d*) 40 wk egg weight

Q. 6. The heritability estimate obtained by the full-sib correlation methods gives

(*a*) Biased estimate (*b*) Larger estimate

(*c*) Estimate with a large sampling error (*d*) Smaller estimate

Q. 7. This is not a metric trait in white leghorn chickens

(*a*) Annual egg number (*b*) Egg weight

(*c*) Skin colour (*d*) Growth rate

Q. 8. The term quantitative genetics

(*a*) Is also called population genetic

(*b*) Refers to the study of quantitative differences in experimental populations

(*c*) Refers to the study of quantitative differences in natural populations

(*d*) Means biometrical genetics

Q. 9. When there is no dominance variance the population means will be reduced to

(*a*) Zero (*b*) a (1-2q)

(*c*) a $(1\text{-}2q)^2$ (*d*) a (p-q)+2dpq

Q. 10. Breeding value of an individual can be known from the

(*a*) Mean value of its progeny

(*b*) Mean value of the population from which it comes

(*c*) Deviation of its progeny from the population mean

(*d*) Can never be known

Q. 11. Avian species that has the highest early growth rate

(*a*) Goose (*b*) Broiler chicken

(*c*) Turkey (*d*) Duck

Q. 12. Genetic variance would be maximum in a

(*a*) Mixed population

(*b*) Inbred population

(*c*) Selected population

(*d*) Isolated population

Q. 13. Additive variance comprises of

(*a*) Genetic variance

(*b*) Variance of breeding value

(*c*) Variance of phenotypic values

(*d*) Dominance variance

Q. 14. In the absence of dominance, dominance variance will be

(*a*) Zero (*b*) One

(*c*) Large (*d*) Small

Q. 15. The genetic variance will be maximal when the gene frequency

(*a*) 0.5 (*b*) 0.1

(*c*) 0.9 (*d*) 1.0-Duplicated

Q. 16. Intangible variation means

(*a*) Environmental variance

(*b*) Non-genetic variance

(*c*) Variance due to errors in measurement

(*d*) Variance whose cause is not known

Q. 17. When repeatability of a trait is high it means

(*a*) Specific environmental variance is least

(*b*) Specific environmental variance is large

(*c*) General environmental variance is large

(*d*) General environmental variance is least

Q. 18. Factor that can increase the heritability estimates

(*a*) High environmental variance

(*b*) Poor environment

(*c*) Variable feeding and Management

(*d*) Uniform conditions of feeding and management

Q. 19. Regression of offspring on one parent is a valid measure for

(*a*) Heritability (*b*) Half of heritability

(*c*) One fourth of heritability (*d*) Genetic correlation

Q. 20. The precision of a heritability estimate is known by the

(*a*) Standard error (*b*) Magnitude

(*c*) Method of estimation (*d*) Experimental design

I. D. Multiple Choice Questions (Tick Correct Answer from Choices)

Q. 1. When heritability is of a character is expected to be very low it is better to estimate it by

(*a*) Regression method (*b*) Half-sib method

(*c*) Full-sib method (*d*) None of these

Q. 2. Which of the following statements about genetic correlation is true?

(*a*) A high phenotypic correlation always imply high genetic correlation

(*b*) Low phenotypic correlation means that the genetic correlation is negative

(*c*) Genetic correlation is always less than phenotypic correlation

(*d*) None of these

Q. 3. The chief drawback of progeny testing is

(*a*) It leads to inbreeding in a closed herd

(*b*) It leads to increase in generation interval and hence reduction in progress per unit of time

(*c*) It is difficult to ensure the same level of management to all the progenies and hence the conclusions are likely to be misleading.

(*d*) None of the above.

Q. 4. The study of gene action of a character can be best under taken through

(*a*) Fitting of model for generation means

(*b*) Fitting of generation variance

(*c*) Either (*a*) or (*b*)

(*d*) By joint scaling test.

Q. 5. Consider two pair of alleles R-r, S-s controlling two different characters. In which R is dominant to r but S is not dominant to s. If the first filial generation is raised from a cross between the parents rrSS and RRss, the number of genotypes expected at F_2 is.

(*a*) 4 (*b*) 6

(*c*) 9 (*d*) 16

Q. 6. The correlation between the genetic value (breeding value) of an individual and the phenotypic value of his full-sib is

(*a*) ¼ h (*b*) ¼ h^2

(*c*) ½ h (*d*) None of these

Q. 7. Let the genotypic value of aa, AA and Aa are –a, +a and d respectively. Then condition of over dominance A is:

(*a*) d negative
(*b*) d < –a
(*c*) d> +a
(*d*) Absolute d > +a

Q. 8. The analysis of segregation using standard error is applied

(*a*) To cases where deviation X^2 cannot be determined exactly
(*b*) When the families segregate into two classes only
(*c*) When a single family segregate into two classes only
(*d*) To all the cases where X^2 can be applied.

Q. 9. To which of the following type of family does the formula for linkage X^2 is given by $X_L = (a_1 - a_2 - 3a_3 + 3a_4)^2/9n$

(*a*) Double back-cross
(*b*) Single back-cross
(*c*) F_2
(*d*) None of these

Q. 10. Which of the following statement is true?

(*a*) The estimate of recombination fraction (p) given by product formula is more efficient than ML estimate
(*b*) Product formula sometimes gives more efficient estimate than by ML method.
(*c*) Product formula applies only when viability differences are not present
(*d*) None of these

Q. 11. Inheritance of quantitative differences depend on genes at

(*a*) Only two loci
(*b*) Many loci
(*c*) Individual locus
(*d*) None of the above

Q. 12. How many generations of full-sib mating are required to achieve 0.50 inbreeding coefficients?

(*a*) Two
(*b*) Three
(*c*) Four
(*d*) Five

Q. 13. Continuous variation in a metric trait is due to

(*a*) Variation in additive genotype and environment

(*b*) Genotype and environment

(*c*) Essentially due to environment

(*d*) Only due to the action of polygenes

Q. 14. When two pair of alleles is segregating independently the phenotypic ratio is

(*a*) Is always 9:3:3:1

(*b*) Is sometimes 9:3:3:1

(*c*) Depends upon the dominance relationships among any one of these

(*d*) None of these

Q. 15. The chief cause of genetic correlation between traits

(*a*) Pleiotropy

(*b*) Linkage

(*c*) Natural selection

(*d*) Physiological mechanism of the traits

Q. 16. Character of low heritability in chickens

(*a*) Viability

(*b*) Egg weight

(*c*) Age at sexual maturity

(*d*) 20 week body weight

Q. 17. The genetic and environmental correlations have different signs, it means that the

(*a*) Traits are unrelated

(*b*) Genetic and Environmental sources of variation affect the character through different physiological mechanisms

(*c*) Traits are highly affected by environment

(*d*) Traits can't be improved upon

Q. 18. Review of literature for genotype-environment interaction for egg production in poultry indicates that

(*a*) They are not generally important

(*b*) They are very important

(*c*) They are very significant for the most traits

(*d*) They are significant for egg production

Q. 19. Trait that is not a component of egg production

(*a*) Age at sexual maturity

(*b*) Body weight at sexual maturity

(*c*) Persistency

(*d*) Intensity of lay

Q. 20. Frequency distributions of most metric traits approximate more or less closely to

(*a*) Normal curves
(*b*) Binomial curves

(*c*) Both (*a*) and (*b*)
(*d*) None of the above

Q. 21. A negative regression of sire's crossbred progeny performance on that of purebreds will indicate the presence of

(*a*) Dominance
(*b*) Over dominance

(*c*) Recessiveness
(*d*) Epistasis

(*e*) Sex-linkage

Q. 22. From which of the following components of variation the estimates of heritability represent the true additive genetic variance

(*a*) Sir component
(*b*) Dam component

(*c*) Sire plus Dam component
(*d*) None of these

Answers

I. A. Multiple Choice Questions

1	(*a*)	6	(*b*)	11	(*d*)	16	(*a*)
2	(*c*)	7	(*d*)	12	(*c*)	17	(*d*)
3	(*a*)	8	(*a*)	13	(*b*)	18	(*a*)
4	(*b*)	9	(*b*)	14	(*b*)	19	(*b*)
5	(*d*)	10	(*a*)	15	(*c*)	20	(*d*)

I. B. Multiple Choice Questions

1	(*a*)	6	(*a*)	11	(*c*)	16	(*a*)
2	(*b*)	7	(*b*)	12	(*d*)	17	(*a*)
3	(*b*)	8	(*b*)	13	(*c*)	18	(*c*)
4	(*c*)	9	(*c*)	14	(*d*)	19	(*c*)
5	(*a*)	10	(*c*)	15	(*d*)	20	(*c*)

I. C. Multiple Choice Questions

1	(*b*)	6	(*b*)	11	(*b*)	16	(*d*)
2	(*a*)	7	(*c*)	12	(*a*)	17	(*a*)
3	(*b*)	8	(*d*)	13	(*b*)	18	(*d*)
4	(*c*)	9	(*b*)	14	(*a*)	19	(*b*)
5	(*d*)	10	(*c*)	15	(*a*)	20	(*a*)

I. D. Multiple Choice Questions

1	(*b*)	7	(*c*)	13	(*b*)	19	(*b*)
2	(*d*)	8	(*c*)	14	(*a*)	20	(*a*)
3	(*b*)	9	(*d*)	15	(*a*)	21	(*b*)
4	(*b*)	10	(*d*)	16	(*a*)	22	(*a*)
5	(*c*)	11	(*b*)	17	(*b*)		
6	(*d*)	12	(*b*)	18	(*d*)		

Chapter 4
Selection Methods in Poultry Breeding

I. A. Multiple Choice Questions
(Tick correct answer from choices)

Q. 1. Condition under which inbreeding and crossing are likely to be better means of improvement than selection without inbreeding

(*a*) Phenotypic variance is non additive

(*b*) Most of the genetic variance is additive

(*c*) Genetic variance is partly additive and partly non-additive.

(*d*) Most of the genetic variance of the character is non-additive.

Q. 2. The most interesting part of the changes caused by selection in a population is

(*a*) Change in population mean

(*b*) Increase in the population mean

(*c*) Reduction in the variance of the trait

(*d*) Change in fitness of the population

Q. 3. Response to selection depends upon

(*a*) Heritability and selection differential

(*b*) Heritability, proportion selected and phenotypic standard deviation

(*c*) Genetic variance of the trait

(*d*) Genetic variance of the trait and intensity of selection

Q. 4 Methods open to the breeder to increase the rate of response to selection

(*a*) Increasing the heritability

(*b*) Increasing the intensity of selection

(*c*) Maximizing heritability and reducing the proportion selected

(*d*) Increasing both heritability and the size of the selected population

Q. 5. By the term asymmetry of response is meant

(*a*) Inequality of responses to selection in opposite directions

(*b*) Lack of symmetry in response from one generation to the other

(*c*) Positive response in one generation followed by negative response in the other generation

(*d*) Erratic fluctuation in response over generations

Q. 6. Selection limit means, it is a state where there is no

(*a*) Genetic variation

(*b*) Phenotypic variation

(*c*) Further response to selection

(*d*) Either (*a*) or (*c*)

Q. 7. The long term selection experiment of Clay-ton Morris and Robertson 1957, indicated that after 34 generations, the phenotypic variance of abdominal bristle number had

(*a*) Decreased (*b*) Increased

(*c*) Not decreased (*d*) Green Constant

Q. 8. A proof that genetic variance is present at the selection limit

(*a*) When there is response to reverse selection

(*b*) When the population mean is remains constant

(*c*) When phenotypic variance is present

(*d*) Heritability remains constant

Q. 9. Individual selection is the method of choice in many programmes because

(*a*) It is simple to operate as it is based on showing appearance

(*b*) It brings about maximum response.

(*c*) It brings about rapid response

(*d*) It needs no record keeping

Q. 10. The family selection is the method of choice for traits

(*a*) With high heritability

(*b*) With low heritability

(*c*) That are expressed in only one sex

(*d*) When family size is large

Q. 11. Progeny testing method of selection means

(*a*) Selection of individual on the basis of its offsprings performance

(*b*) Selection of progeny on the basis of their mean phenotypic value

(*c*) Selection of outstanding progeny after testing them.

(*d*) Selection of progeny of a sire after comparing with different sire groups.

Q. 12. Indirect selection can be superior to direct selection when

(*a*) Desired character has high heritability

(*b*) Secondary character has higher heritability

(*c*) Genetic correlation between the traits is high

(*d*) Due to both (*b*) & (*c*)

Q. 13. Which one of the following methods is the most efficient for improving overall economic value of cows?

(*a*) Selection for several traits by using independent culling levels

(*b*) Selection for total score

(*c*) Selection for one trait each time

(*d*) Selection for three traits at a time.

Q. 14. In a programme of reciprocal recurrent selection how many lines are needed to start with

(*a*) 2 (*b*) 8

(*c*) 4 (*d*) 6

Q. 15. In a programme of recurrent selection for specific combining ability what should be the nature of the tester population

(*a*) Similar to selected line

(*b*) Synthetic population

(*c*) Random bred

(*d*) Inbred

Q. 16. Method of improving the accuracy of measurement of response to selection is by

(*a*) Reducing the environmental variance

(*b*) Having a unselected control population

(*c*) Having a selected flock in opposite direction

(*d*) Having a control population

Q. 17. The chief circumstance under which within family selection is preferred

(*a*) Little amount of environmental variance common to members of a family

(*b*) Too long a generation interval

(*c*) Low heritability

(*d*) Large component of environmental variance common to members of a family

Q. 18. By the term sib-selection means

(*a*) Same as family selection

(*b*) Selected sib is only measured

(*c*) Selected sib is not measured

(*d*) All the members measured

Q. 19. The selection technique of great pertinence to poultry improvement is

(*a*) Directional

(*b*) Disruptive

(*c*) Stabilizing, heterozygotes preferred

(*d*) Stabilizing extremes preferred.

Q. 20. Who said "Selection is waste basket category that includes all causes of directed change in gene frequency that do not involve mutation or introduction for without"

(*a*) J. B. S.Haldane (*b*) C. Darwin

(*c*) S. Wright (*d*) Dobzhansky

I. B. Multiple Choice Questions (Tick Correct Answer from Choices)

Q. 1. Selection means choosing of

(*a*) Good individuals

(*b*) Individuals to be used as parents

(*c*) Individuals about the population mean

(*d*) Superior individuals with desirable traits

Q. 2. The term response means

(*a*) Change in the phenotypic value

(*b*) Change in the genetic value

(*c*) Superiority of the selected parents

(*d*) Difference of the mean phenotypic value between the offspring and base population

Q. 3. Intensity of selection can also be stated as

(a) Standardized selection differential

(b) S/ genetic standard deviation

(c) R/ phenotypic standard deviation

(d) H/ standard deviation of breeding values

Q. 4. Individual selection means

(a) Selection of the progeny on the basis of individual's phenotype

(b) Selection of individuals on the basis of average performance of half-sibs

(c) Selection of animals on the basis of their phenotypic values

(d) Selection of individuals on the basis of their genetic values.

Q. 5. Family selection means selection of

(a) Parents on the basis of their offsprings' average performance

(b) Family members on the basis of the phenotypic value of sire

(c) The best animal of the family on the basis of its deviation from the family mean.

(d) A family on the basis of the deviation of the family mean from the population mean

Q. 6. The scientist who had first compared family merit and individual merit as based on selection in pigs

(a) J. L. Lush (b) R. Osborne

(c) H. F. Smith (d) R. A. fisher

Q. 7. Family selection tends to be superior to individual selection when

(a) Heritability of family means is larger than that of individual values

(b) Standard deviation of family means are less than those of individual values (*c*) The correlations between family members is high

(*d*) There is considerable variation due to common environment

Q. 8. Selection scheme followed in poultry breeding projects in our country for improvement of egg production

(*a*) Individual selection

(*b*) Family selection

(*c*) Combined selection

(*d*) Selection of half-sib families

Q. 9. Who was the author of the publication? The efficiency of three methods of selection

(*a*) Hazel & Lush (*b*) Smith, F.

(*c*) Wright, S. (*d*) Osborne, R.

Q. 10. In a programme of reciprocal recurrent selection how many lines are needed to start with

(*a*) 2 (*b*) 8

(*c*) 4 (*d*) 6

Q. 11. What is the essential pre-requisite in the lines used for reciprocal recurrent selection?

(*a*) Inbred

(*b*) Random bred

(*c*) Difference in gene frequency for the trait

(*d*) Genetically similar & superior

Q. 12. Scientists who assessed theoretically the merits of recurrent reciprocal selection

(*a*) Lerner & Cruden

(*b*) Hazel & Lush

(*c*) Comstock, Robinson & Harvey

(*d*) Falconer & Robertson

Q. 13. Another name for intensity of selection

(*a*) Expected selection differential

(*b*) Standardized selection differential

(*c*) Weighted selection differential

(*d*) Realized selection differential

Q. 14. Theoretically, Inbreeding followed by crossing of the inbred lines in a large population without any selection would result in

(*a*) No change in the population mean

(*b*) Increase in the population mean

(*c*) Decrease in the population mean

(*d*) We can't predict the outcome

Q. 15. Long term selection experiments in Drosophila indicate that in long term selection as the selection limit is approached the phenotypic variance tends to

(*a*) Increase

(*b*) Decrease

(*c*) Remain almost constant

(*d*) To show no trend at all

Q. 16. The resemblance between offspring and parents provides the basis for

(*a*) Discrete breeding

(*b*) Inbreeding

(*c*) Random breeding

(*d*) Selective breeding

Q. 17. The ratio of response to selection to the selection differential is the

(*a*) Degree of genetic determination

(*b*) Heritability in broad sense

(*c*) Realized heritability

(*d*) Repeatability

Q. 18. Mitosis results in

(*a*) Reduction of chromosome numbers

(*b*) Doubling of chromosome numbers

(*c*) No change in chromosome numbers

(*d*) None of the above

I. C. Multiple Choice Questions (Tick Correct Answer from Choices)

Q. 1. Variation in the generation mean for the estimation of response can be overcome by the use of

(*a*) Control population

(*b*) Use of several lines

(*c*) By increasing the intensity of selection

(*d*) None

Q. 2. Response in both the lines in two way selection experiment is usually

(*a*) Equal

(*b*) More in up line

(*c*) More in down line

(*d*) Does not follow any trend

Q. 3. Selection limit represents the

(*a*) Fixation of all favorable alleles

(*b*) All alleles at intermediate frequencies

(*c*) Aggregation of all dominant genes

(*d*) None

Q. 4. If the population after reaching the selection limit responds to reverse selection, it may be due to

(*a*) Environmental effect

(*b*) Conflict between natural and artificial selection

(*c*) Pleotrophic gene action

(*d*) None of these

Q. 5. Selection based on the performance of the individual, its full-sibs and half-sibs is known as

(*a*) Family selection (*b*) Sire family selection

(*c*) Combined selection (*d*) None

Q. 6. In family selection, if the minimum number of parents fixed is ten, then at least_______must be selected in every cycle of selection

(*a*) Ten (*b*) Twenty

(*c*) Five (*d*) None

Q. 7. If selection is practiced for the ovarian response to gonadotropic hormone in mice, which method of selection should be used?

(*a*) Mass selection (*b*) Sib-selection

(*c*) Family selection (*d*) None of these

Q. 8. The correlation in the breeding values among full-sibs is

(*a*) 1/4 (*b*) 1/2

(*c*) 1/8 (*d*) None of these

Q. 9. The correlation in the breeding values among half-sibs is

(*a*) 1/4 (*b*) 1/2

(*c*) 1/8 (*d*) None of these

Q. 10. The relative merit of individual selection in comparison to that of combined selection is greatest when the value of intra-class correlation is

(*a*) 0.5 (*b*) 0.25

(*c*) 0.75 (*d*) None of these

Q. 11. The genetic and environmental causes of correlation will be

(*a*) Always of the same sign

(*b*) Always of the same magnitude

(*c*) Always of different signs

(*d*) May be of the same and different signs.

Q. 12. The indirect selection can not be expected to be superior to direct selection unless the secondary character has______heritability than the desired character

(*a*) Substantially higher (*b*) Substantially lower

(*c*) Equal (*d*) None of these

Q. 13. One of the circumstances most likely to render indirect selection superior to direct selection is_______

(*a*) Difficulty in measure of the traits with precision

(*b*) Very low heritability

(*c*) Large component of environmental variance

(*d*) None of these

Q. 14. Multi-stage selection may be practiced for_______

(*a*) Reducing the breeding cost

(*b*) More effectiveness

(*c*) Simplicity in application

(*d*) None of these

Q. 15. For the selection of individuals at early ages, the basis of selection should be_______

(*a*) Sib performance (*b*) Parents performance

(*c*) Grand parent performance (*d*) None of these

Q. 16. The alternate method to the use of control population to measure the genetic response is

(*a*) Sib selection

(*b*) Family selection

(*c*) Divergent selection

(*d*) Individual selection

Q. 17. For estimation of variance components of all the variables, the most suitable model for analysis of data is

(*a*) Fixed model (*b*) Mixed model

(*c*) Random model (*d*) All the above

Q. 18. Random sample test for layers and broilers provides valuable information about

(*a*) Ranking of different genetic stocks

(*b*) Commercial viability of the stocks

(*c*) Genetic progress achieved in their performances over time

(*d*) All the above

Q. 19. Regression of the offspring on mid parental value is equal to the

(*a*) Heritability

(*b*) Selection differential

(*c*) Half of the heritability

(*d*) Double of the heritability

Q. 20. Mean dominance deviation in a population is equal to

(*a*) Average mean breeding value

(*b*) Average genotypic value

(*c*) One

(*d*) Zero

I. D. Multiple Choice Questions (Tick Correct Answer from Choices)

Q. 1. Rate of response to selection can not be increased by

(*a*) Increasing heritability

(*b*) Reducing environmental variance

(*c*) Increasing environmental variance

(*d*) Standardizing the management

Q. 2. The prediction of response is generally valid for

(*a*) One generation

(*b*) Two generation

(*c*) Several generations

(*d*) Not even for one generation

Q. 3. Individual merit and family merit as basis of selection was proposed by

(*a*) Hazel and Lush (1942) (*b*) Hazel (1943)

(*c*) Lush (1947) (*d*) Lerner (1952)

Q. 4. How many models were given by Griffing in diallel analysis?

(*a*) Two (*b*) Three

(*c*) Four (*d*) Five

Q. 5. Which of the following is a particular case of combined selection with equal weights assigned to between family and within family components?

(*a*) Mass selection (*b*) Family selection

(*c*) Within family selection (*d*) None of these

Q. 6. Which of the following do not fall into the categories of simultaneous selection for multiple traits?

(*a*) Combined selection

(*b*) Selection by discriminant function

(*c*) Independent culling levels

(*d*) None of the above

Q. 7. The purpose of constructing a selection index is

(*a*) Economic improvement of a set of individuals through account selection

(*b*) Improvement through better intensity of selection

(*c*) Improve economic value by manipulating both the above

(*d*) None of these

Q. 8. The method of selection which cannot be applied in the case of sex-limited trait is

(*a*) Individual selection (*b*) Sib selection

(*c*) Progeny testing (*d*) Pedigree selection

Q. 9. For achieving genetic improvement in a character by selection based on another character.

(*a*) The characters should be genetically correlated

(*b*) They should be positively genetically correlated

(*c*) Their heritability should be very high

(*d*) Both characters should be measurable

Q. 10. Most appropriate method of breeding for a population with a large amount of additive genetic variance is

(*a*) Cross breeding

(*b*) Selective breeding

(*c*) Inbreeding

(*d*) Line breeding

Q. 11. The objectives of control population are

(*a*) To assess the magnitude of short term fluctuations in the environment.

(*b*) To maintain genetic constancy over a long period

(*c*) To serve as gene pools and stable base population

(*d*) All the above

Q. 12. Un-pedigreed random breeding control is one that has

(*a*) A random breeding population without selection

(*b*) Carefully sampled from a base population

(*c*) Both

(*d*) None

Q. 13. A. I. in un-pedigreed random breeding control population can be done with the objective that

(*a*) Each male is mated with same number of females

(*b*) To reduce the variation in number of progeny per parent

(*c*) Most of the parents contribute progeny to the next generation

(*d*) All of the above

Q. 14. Pedigreed random breeding control population is one that

(*a*) Each individual is reproduced each generation with full pedigree information

(*b*) Sampling among individuals is eliminated

(*c*) The ratio of males and females in the population is close to 1:1

(*d*) All the above

Q. 15. The control population can be

(*a*) Un-pedigreed random breeding control population

(*b*) Pedigreed random breeding control population

(*c*) Repeat mating control population

(*d*) All the above

Q. 16. The genetic correlation between part and annual egg production is______

(*a*) High (*b*) Low

(*c*) Inconsistent (*d*) Moderate

Q. 17. Full sib family selection is preferred when reproductive rate is

(*a*) High (*b*) Low

(*c*) Not specific (*d*) None of these

Q. 18. It has been mostly observed that combined selection with optimum weights attached to the component traits resulted in______genetic progress than either individual or family selection.

(*a*) Less (*b*) More

(*c*) Consistent (*d*) Inconsistent

Q. 19. The threshold characters follow______distribution

(*a*) Binomial

(*b*) Normal distribution

(*c*) Positively skewed distribution

(*d*) Negatively skewed distribution

Q. 20. For the improvement of the traits with high heritability _______method of selection is used.

(*a*) Individual selection (*b*) Combined selection

(*c*) Progeny selection (*d*) None

Q. 21. To utilize additive gene effects_______selection method is used.

(*a*) Reciprocal recurrent selection

(*b*) Intra-population selection

(*c*) Recurrent selection

(*d*) None.

Q. 22. If the response is divided by_______it can be used to compare the response in the traits expressed in different units

(*a*) Genetic standard deviation

(*b*) Selection differential

(*c*) Phenotypic standard deviation

(*d*) None

Q. 23. Intensity of selection is proportional to

(*a*) Percentage of individuals falling within the point of truncation

(*b*) Percentage of individuals selected

(*c*) Method of selection

(*d*) None

Q. 24. The response of selection can be predicted if

(*a*) Heritability is known

(*b*) Heritability and selection intensity are known

(*c*) Heritability, selection intensity and phenotypic standard deviation are known

(*d*) Selection intensity and selection differential are known

II. Fill in the Blanks

Q. 1. Mention the most important traits need to be improved in

(*a*) Broiler parents

(*i*) _______

(*ii*) _______

(*iii*) _______

(*b*) Broiler progeny

(*iv*) _______

(*v*) _______

Q. 2. Mention three important traits need to be improved in layer stock

(*vi*) _______

(*vii*) _______

(*viii*) _______

III. State Following Statements if True or False

1. Selection is the most important force available to the breeder for changing the gene frequencies.

2. In tropics, selective breeding is most effective where the traits have lower additive genetic variance.

3. In the method of reciprocal selection for improving the egg production, though the performance of the cross is expected to increase, the performance of the line themselves in respect of characters selected is expected to decrease.

4. Recurrent selection is one of the important methods of selection to improve both general and specific combining ability of two specific populations regardless of the level of dominance.

5. In general, selection for meat production is more complex than selection for egg production.

6. Giving the value to a sire on the basis of the performance of the relatives is known as sire index.

7. If the heritability differs between populations the responses to selection remains to be the same.

8. An unselected control population will reveal how much asymmetry can be attributed.

9. The response to selection cannot be expected to continue indefinitely and sooner or later all the alleles segregating will be brought to fixation.

10. The use of information from relatives is not of great importance in the application of selection to animal breeding.

11. When the intensity of selection is taken into consideration, the optimum family size gives the greatest expected response.

12. The response of a correlated character can not be predicted if the genetic correlations and the heritability of the two characters are known.

13. Close agreement between observed and predicted correlated response can not always be expected if the genetic correlation is low.

14. Natural selection takes no account of heritabilities or genetic correlation and is therefore, less efficient in improving fitness characters than artificial selection.

15. Mother's theory of polygenic balance is based on the idea of selection favouring intermediate values of metric character, the effect this is likely to have on linkage.

16. Selection differential means average superiority of the selected parents in the population.

17. R.A Fisher compared family merit and individual merit as basis of selection in pigs.

18. The resemblance between offspring and parents provides the basis for selective breeding.

19. The prediction of response is generally valid for one generation.

20. The ratio of response to selection to the selection differential is repeatability.

21. Progeny testing is the most valuable aid for selection of traits of high heritability.

22. In random mating populations, the zygotic ratio is the square of the gametic ratio.

23. Significant differences for egg production among populations reared under identical conditions of management and feeding is an indication of genetic differences among them.

24. Sex of the progeny is usually determined by chromosome compliments of the male chicken.

25. Egg production in poultry is a sex limited trait.

26. When the commercial broiler is a product of 3-way cross the female parent is usually a single cross.

27. Heterosis is twice the inbreeding depression in opposite direction.

28. More than 95% of the commercial broilers and layers sold in the market today are crosses of one kind or other.

29. Simultaneous selection for several characters is superior to selection for single trait at a time to improve the overall merit of the population under selection.

Answers

I. A. Multiple Choice Questions

1.	(*d*)	6.	(*d*)	11.	(*a*)	16.	(*c*)
2.	(*a*)	7.	(*a*)	12.	(*b*)	17.	(*d*)
3.	(*d*)	8.	(*a*)	13.	(*b*)	18.	(*c*)
4.	(*c*)	9.	(*b*)	14.	(*a*)	19.	(*a*)
5.	(*a*)	10.	(*c*)	15.	(*d*)	20.	(*c*)

I. B. Multiple Choice Questions

1.	(*b*)	6.	(*c*)	11.	(*b*)	16.	(*c*)
2.	(*d*)	7.	(*d*)	12.	(*a*)	17.	(*d*)
3.	(*d*)	8.	(*b*)	13.	(*c*)	18.	(*d*)
4.	(*a*)	9.	(*c*)	14.	(*d*)	19.	(*c*)
5.	(*a*)	10.	(*b*)	15.	(*c*)	20.	(*c*)

I. C. Multiple Choice Questions

1.	(*a*)	6.	(*a*)	11.	(*d*)	16.	(*c*)
2.	(*b*)	7.	(*b*)	12.	(*a*)	17.	(*c*)
3.	(*a*)	8.	(*b*)	13.	(*a*)	18.	(*d*)
4.	(*b*)	9.	(*a*)	14.	(*a*)	19.	(*a*)
5.	(*c*)	10.	(*a*)	15.	(*b*)	20.	(*d*)

I. D. Multiple Choice Questions

1.	(*c*)	7.	(*a*)	13.	(*d*)	19.	(*a*)
2.	(*a*)	8.	(*a*)	14.	(*d*)	20.	(*a*)
3.	(*b*)	9.	(*b*)	15.	(*d*)	21.	(*b*)
4.	(*c*)	10.	(*b*)	16.	(*a*)	22.	(*c*)
5.	(*a*)	11.	(*d*)	17.	(*a*)	23.	(*a*)
6.	(*c*)	12.	(*c*)	18.	(*b*)	24.	(*c*)

II. Fill in the Blanks

(*i*) Egg production

(*ii*) Hatchability

(*iii*) Fertility

(*iv*) Meat production or body weight

(*v*) Feed consumption

(*vi*) Egg production or persistency

(*vii*) Early maturity

(*viii*) Egg size

III. True or False

1.	True	8.	True	15.	True	22.	True
2.	False	9.	True	16.	True	23.	True
3.	False	10.	False	17.	True	24.	False
4.	False	11.	True	18.	True	25.	True
5.	True	12.	False	19.	True	26.	True
6.	True	13.	True	20.	False	27.	False
7.	False	14.	True	21.	False	28.	True
						29.	True

Chapter 5

Poultry Breeding Plan

I. Fill in the Blanks

1. The realized genetic correlations between body weight and egg production obtained from correlated response studies in selection experiments are mostly reported to be_______in direction.

2. White Cornish is always used as_______line parent for better broiler production.

3. During the course of developing inbred lines most of the lines become extinct because of_______and_______.

4. The estimates of genetic variance and co-variance are generally expressed as :

   (*a*) _______

   (*b*) _______

   (*c*) _______

5. The various types of gene effects can be measured in terms of :

   (a) ______

   (b) ______

   (c) ______

6. A high correlation between purebred and crossbred paternal half-sibs performances indicates the importance of______gene action.

7. If the sire x breed of dam interaction component for a trait is significant, it indicates the presence of______

8. The selection methods utilizing the family averages as the criterion are more successful under the following conditions:

   (a) ______

   (b) ______

   (c) ______

9. Sire family selection is more efficient than dam family selection when the heritability of the trait is less than______

10. Combined selection takes into account______and family averages for the traits of low heritability.

11. Combined selection as a method of selection for traits of low heritability in poultry has been discussed in great detail by______(Give the name of the scientist)

12. The efficiencies of different methods of selection such as mass selection, sire family selection, dam family selection and index selection when applied to same base population has been reported by______(Give the name / names of the scientist).

13. The method of selection index was originally developed for application to animal improvement by______(Give the name / names of the scientist).

14. The variation due to reciprocal effects are due to_______

15. Specific combining ability involves the following gene actions

    (*a*) _______

    (*b*) _______

    (*c*) _______

16. The amount of inbreeding co-efficient achieved after 3 (three) generations of full-sib mating is_______%.

II. Fill up the Blanks with a Suitable Word Given in the Parenthesis

1. The genetic correlation between part and annual egg production is_______(High, Low, Inconsistent)

2. The negative regression of sire's crossbred progeny performance on that of purebreds will indicate the presence of_______(dominance, over dominance, recessiveness, epistasis, sex-linkage)

3. From which of the following components of variation the estimates of heritability represent the true additive genetic variance_______(Sire component, Dam component, Sire + dam component)

4. Full-sib family selection is preferred when reproductive rate is_______(High, Low, not specific)

5. It has been mostly observed that combined selection with optimum weights attached to the component traits resulted in_______genetic progress than either individual or family selection. (Less, more, inconsistent).

6. When the heritability of a trait is high, it is advisable to adopt_______selection for its improvement. (Individual, pedigree, family selection).

III. State True or False for the Following Statements

1. The same criteria of selection should be used for both male and female lines parents for improving the broiler production.
2. There is differential contribution of sire and dam to the profit from broiler enterprise from breeding point of view.
3. Review of literature indicates that broiler weight is a highly heritable trait.
4. The estimates of heritability for viability has been invariable found more than that of egg production.
5. Intra-population selection methods appear to be more suitable for improving the viability in chicken.
6. Usually negative genetic correlation exists between the egg production and body weight.
7. Breeds and strains crosses have been almost always found to be superior to purebred for broiler production.
8. The strain crosses within a breed have been mostly observed to be superior to the strain crosses between the breeds.
9. The development of strains within a breed is less costly compared to development of inbred lines.
10. It has been observed that there is a differential response of lines to inbreeding for egg production, hatchability, fertility and viability.
11. The regression and correlation will be the same when the variances within the purebred and crossbred progenies are same.
12. If gene frequencies in the segregating populations are at equilibrium no improvement is expected in a selection programme based on crossbred performance.

13. Inter-population selection may be defined as the selection of individuals within a population to improve combining ability.

14. Intra-population selection methods mainly utilize non-additive genetic variances.

15. Mass selection is preferred when the heritability and environmental variance of the trait is high.

16. The heritability of egg production ranges from moderate to high in magnitude.

17. Before selecting top a few families to produce next generation, each family average is required to be weighed according to the number of individuals within a family.

18. In progeny testing the progress per unit of time is more than the progress per generation.

19. The efficiencies of indices increase when information is combined from more sources in a combined selection programme.

20. Within family selection is not practiced for improving egg production in poultry.

21. The response obtained from selection is a function of selection differential and genetic correlation derived from sire component of variance and co-variance.

22. It has been observed that selection limit is least depressed when selection limit is no greater than half.

23. When selection limit is reached for a trait, no improvement is possible by increasing the intensity of selection.

24. The selection differential achieved becomes smaller when several characters are considered simultaneously than tandem selection.

25. The principle involved in constructing a linear selection index is not essentially to maximize the correlation between the index and aggregate genotype.

26. If the index is to be used for selection, the weights to be attached to the different characters should be obtainable in a short interval of time.

27. The performance of 3-way and 4-way crosses can be predicted on the basis of the performance of the constituent single crosses.

28. When the inbreeding co-efficient is unity the variance between crosses is the sum of all the components of the genetic variance in the base population.

29. The amount of heterosis following a cross between two particular lines or population depends on the square of the difference of gene frequency between the populations.

30. In sib-selection, the selected individuals also contributed to the estimate of their family mean.

31. The loci without dominance neither cause inbreeding depression nor heterosis.

32. The cause of variation between the generation means is only environmental in a finite population.

33. The amount of heterosis produced in F_2 is ¼ of that produced in F_1 which is produced by mating two different populations.

34. In the bidirectional selection experiments high line serves as an environmental control for the low line but the *vice-versa* is not true.

35. When the population is under artificial selection and then relaxed, the natural selection acts uni-directionally.

36. The main advantage of 3-way or 4-way cross is to overcome the reproductive disadvantage.

37. The birds under the random sample tests at one station are kept under the diversified conditions in order to suit the different agro-climatic conditions existing in our country.

Answers

I. Fill in the Blanks

1. Negative
2. Male Line
3. Viability and reproductive fitness
4. (a,b,c)—Phenotypic standard deviations, heritability, Phenotypic and genetic correlations.
5. (a,b,c)–General combining ability, specific combining ability, reciprocal effects.
6. Additive
7. Heterosis
8. (a,b,c)–Large family size, low heritability, high environmental variance.
9. 0.1.
10. Individual's average
11. Osborne (1957a,b)
12. Kinney *et al.* (1970)
13. Hazel and Lush (1942)
14. Maternal effect and sex linked genes
15. Dominance, over-dominance and epistasis
16. 50%

II. Fill in the Blanks with a Suitable Word

1. High
2. Over-dominance
3. Sire component
4. High
5. More
6. Individual

III. True or False

1.	False	11.	True	21.	False	31.	True
2.	True	12.	True	22.	True	32.	False
3.	True	13.	False	23.	True	33.	False
4.	False	14.	False	24.	True	34.	False
5.	False	15.	False	25.	False	35.	True
6.	True	16.	False	26.	True	36.	True
7.	True	17.	True	27.	True	37.	False
8.	False	18.	False	28.	True		
9.	True	19.	True	29.	True		
10.	True	20.	True	30.	False		

Chapter 6

Advanced Poultry Breeding

I. A. Fill in the Blanks

1. Basic effect of selection is (*a*)_______causing observable change (*b*)_______

2. Natural selection operates through differences. (*a*)_______or (*b*)_______

3. Response to selection symbolized by R is the difference of mean_______between the offspring of the selected parents and the whole of the_______before selection.

4. Selection differential symbolized by S is the mean_______value of the individuals selected as parents expressed as a_______from the_______

5. The connection between the response (R) and the selection differential (S) is given as_______

6. Regression of offspring on mid parent is equal to the heritability provided there is no_______cause of resemblance between_______and_______and there should be no natural selection.

7. Standardized selection differential is called______

8. Expected Response R=______

9. Improvement of Response can be done by increasing (*a*) ______by reducing environmental variation through better and uniform management and by increasing (*b*)______

10. The response to selection can be used for estimating heritability in the______population if R and S are known. The heritability estimated in this way is called the______heritability.

11. In bidirectional selection experiments, the response is symmetrical or asymmetrical.

12. Who developed theory of the sampling variance of realized heritabilities and gave formulae to calculate the approximate standard error when there is only one selection line.

13. When selection limit is reached, all relevant loci become______either by selection or by the accompanying______

14. While selecting individuals on family average basis, under what circumstances out of the two individuals with equal phenotypic values; one from good family is preferred over the other individual belonging to a bad family and *vice-versa*.

15. The phenotypic value of an individual measured as a deviation from the population mean is the sum of (1) deviation of its family mean (pf) from the population mean and (2) deviation of the individual from the family mean (pw). Hence, P = Pf + Pw. What is the weightage given to these two components in (*a*) individual selection, (*b*) family selection and (*c*) within family selection?

    (*a*)______ (*b*)______

    (*c*) ______

16. If over dominance is more important for a particular character which of the two systems is more appropriate for obtaining better results.

    (*a*) Crossing the inbred lines.

    (*b*) Crossing strains with little or no inbreeding.

17. What are the causes of genetic correlations between any two characters (*a*)_______(*b*)_______

18. The response of a correlated character can be predicted if the_______and the_______of the two characters are known.

19. The improvement of performance in an environment if different from the one in which selection was carried out can therefore, be predicted from a knowledge of the_______of performance in each environment and the_______between the two performances.

20. In multiple trait selection appropriate weightage is given to each character according to its_______, _______and the_______and_______correlations between the different characters.

21. For improving egg number without deteriorating egg weight, which selection method is advisable? Considering egg number only under family selection or index selection or multiple traits selection including egg number and egg weight both.

22. If selections are based on part time egg production only, how much response is expected per unit of time as compared to, if selection were based on annual egg production.

I. B. Fill in the Blanks

1. The effect of selection at the genetic level is to change the_______of the population.

2. The proportionate contribution of offspring to the next generation is called______of the individual.

3. Effect of selection on gene frequency depends upon

   (*a*) ______

   (*b*) ______

4. Natural selection operates essentially by differential______

5. There are two ways in which the action of the breeder can change the genetic properties of a population. They are:

   (*a*) ______

   (*b*) ______

6. If there are no differences in fertility among the selected individuals or of viability among their progeny, then the gene frequency are the______in the offspring generation as in the selected parents.

7. Difference of mean phenotypic value between the offspring of selected parents and whole of parental population before selection is known as______

8. The deviation of the mean phenotypic value of the individuals selected as parents from the population mean is known as______

9. The response to selection for a metric trait is the function of______and______

10. The standardized selection differential is known as______

11. The ratio between response to selection and selection differential is known as______

12. If the selection is to be carried out for a long term, the total range is likely to be between______times the square root of the additive genetic variance.

13. When the correlation between genotype and phenotype of a trait is high______selection method is more effective and simple to operate.

14. If the heritability of a trait is 0.20 or less______selection method is more effective.

15. If the component of environmental variance common to members of a family is large______selection method is more effective.

16. When the criterion of selection is to bring about improvement in the aggregate genetic economic gain______method is used.

17. In Q. No. 16 if a particular trait is required to be kept at the genetic equilibrium______selection method is used.

18. The gene action utilized in intrapopulation selection method is______in nature.

19. The observable components of between cross variance are: (*a*) ______ (*b*) ______

20. There are two components of reciprocal variance. They are: (*a*) ______ (*b*) ______

21. For the improvement of the traits with high heritability______method of selection is used.

II. A. State if True or False for the Following Statements

1. Inbreeding followed by crossing alone cannot be expected to lead to improvement of the poultry population unless supplemented by selection.

2. Crossing of inbred lines produces no genotypes that could not occur in the base population.

3. The condition under which inbreeding and crossing are likely to be better means of improvement than selection without inbreeding is when much of the genetic variance of the character is non-additive.

4. If the lines crossed are fully inbred there will be no genetic variance within crosses and the variance between crosses will be equal to the genotype variance in the base population.

5. There is some degree of correlation between a line performance as an inbred and its general and specific combining ability so as proportion of lines can be discarded on the basis of their own performance before the crosses are made.

6. Over-dominance is the property shown by two alleles when the heterozygote lies in between the range of the two homozygotes.

7. If matings are done between full-sib in a poultry flock for several generations the genetic variability would be reduced.

8. If you know the inbreeding coefficient of an individual you can know for certain how many loci are homozygous in that individual.

9. Effective flock size is always smaller than the actual flock size.

10. Inbreeding as a random process, affects all loci, and the mean values of all characters showing directional dominance decline. But under RRS it is only the selected character that should decline except in so far as linked loci are carried along.

11. Inbreeding causes an increase in the frequency of the homozygous genotypes and a decrease of heterozygous genotypes in a poultry population.

12. Genes at intermediate frequencies contribute more to a change of mean than genes at high or low frequencies.

13. The change in the population mean due to inbreeding is almost linear when plotted against the inbreeding coefficients.

14. Fitness traits are generally least affected due to inbreeding.

15. Loci without dominance cause neither inbreeding depression nor heterosis.

16. Inbreeding causes genetic differentiation between lines and genetic uniformity within lines.

17. Inbred lines show less variation due to environmental effects than random bred flocks.

18. When inbreeding is complete in a flock the genetic variance in the population would be almost double.

19. During the progress of inbreeding in a flock the contribution of additive variance to the total between cross variances increases linearly with the inbreeding coefficient.

20. When a set of inbred lines are crossed at random we can compute for each line its mean cross performance which is called specific combining ability.

21. Reciprocal recurrent selection is a method for the simultaneous improvement of both general and specific combining ability of the lines.

22. An essential pre-requisite in the selection for lines for RRS programme is that there should be some difference in the gene frequency for the trait under selection between the two lines.

23. Layer type chickens are either inbred-hybrids or strain crosses whereas commercial broilers are usually breed crosses.

24. According to the Hardy Weinberg law in a large population if any one individual of one sex has free choice to mate with any individual of the opposite sex, it is called random mating.

25. Pan-mixia and random mating are synonyms.

26. Hardy Weinberg law was discovered by a team of scientists Hardy and Weinberg (1908) who were working together in the field of population genetics.

27. Complete positive assortative mating should lead to complete homozygous of a population very quickly.

28. Assortative mating based on external resemblance may lead to a population with genetic composition very different from that reached by inbreeding based on relationship.

II. B. State the Following Statements True or False

1. Genes at intermediate frequency do not change by selection.

2. The effect of selection on gene frequency depends on the intensity of selection and the initial frequency of gene/ genes have no role to play in it.

3. Natural selection refers to the differential rate of reproduction.

4. When the environmental variation is reduced the response obtained from the selection is increased.

5. When the phenotypic standard deviation of a metric trait is decreased the magnitude of selection differential is also reduced.

6. Usually the progress per unit of time assumes more importance than progress per generation in a selection programme.

7. Weighting of the selection differential by the number of offspring produced by each parent or pair of parents takes account of effects of natural selection.

8. If the value of effective selection differential is less than that of the expected selection differential, it shows that the natural selection is operative against the selection.

9. One of the reasons of asymmetry of response to selection in opposite direction is epistasis.

10. In a long term selection experiment, the total range may be only about 4 times the phenotypic standard deviation.

11. In family selection the within family deviation is given zero weight.

12. In sib selection, the selected individuals do not contribute to the estimate of their family mean.

13. In within family selection, every family contributes equally to the parents of the next generation.

14. Genetic slippage may be one of the reasons of refractory response.

15. The higher interactions contributing to the total between cross variance have coefficients in correspondingly higher powers of F (inbreeding coefficient).

16. The aggregate genetic economic worth of an individual is the aggregate of phenotypic value of each trait multiplied with corresponding economic value.

17. For the construction of selection index, the values of regression coefficient of each trait are so chosen that the correlations between the index and the aggregate genetic economic value remain high.

18. In independent culling level method of selection the individual gets premium for its out-standing merit in one trait to offset its deficiency in another trait.

19. Sometimes it is possible to achieve more rapid progress in a correlated character than one for which selection is made.

20. In order to check the improvement of a trait in undesirable direction during index selection, the selection may be performed in two stages.

21. A number of experiments have shown that the effectiveness of independent culling level lies in between individual selection and index selection.

22. The positive genetic correlation between part and full record selection may turn out to be negative after several cycles of part record selection.

23. In reciprocal recurrent selection general and specific combining ability are utilized.

24. When the reproduction rate is low, sire family selection should be used.

25. Even if the intensity of selection is not same, the realized heritability can be used to make comparison of the effectiveness of different selection experiments.

26. Fisher (1936) was the first person who applied the index selection for animal improvement.

27. If the covariance between sire's purebred and crossbred progeny is negative, it indicates involvement of over dominance.

28. Accuracy of family selection increases with the increase of family size.

29. Progeny selection brings about more improvement per unit of time though per generation improvement is slow.

Answers

I. A. Fill in the Blanks

1. (*a*) To change gene frequency causing observable change in (*b*) population mean

2. Natural selection operates through differences (*a*) in fertility among the parents or (*b*) of viability among their progeny.

3. Response to selection symbolized by R is the difference
o f
mean phenotypic value between the offspring of the selected parents and the whole of the parental generation before selection.

4. Selection differential symbolized by S is the mean phenotypic value of the individuals selected as parents-expressed as a deviation from the population mean.

5. The connection is between the response (R) and the selection differential (S) is given as
$R = bop\ S$
bop = Regression of offspring on mid parent.

6. Regression of offspring on mid parent is equal to the heritability provided there is no non-genetic cause of resemblance between offspring and parents and there should be no natural selection.

7. Standardized selection differential $S/\sigma\ p$ is called the intensity of selection.

8. Expected Response $R = ih^2 \sigma P$

9. Improvement of response can be done by increasing (*a*) Heritability by reducing environmental variation through better and uniform management and by increasing (*b*) intensity of selection.
10. The response to selection can be used for estimating heritability in the base populations if R and S are known. Heritability estimated in this way is called the Realized heritability.
11. Asymmetrical
12. Hill (1973, 1972c, d, 1977).
13. When selection limit is reached all relevant loci become homozygous either by selection or by accompanying inbreeding.
14. When the differences between families are mainly genetic and the *vice-versa* when the differences are mainly environmental.
15. (*a*) Equal (*b*) Pf=1, Pw (0) (*c*) Pf = 0, Pw =1
16. Crossing the inbred lines.
17. (*a*) Pleotrophic effect (*b*) Close linkage
18. The response of a correlated character can be predicted if the genetic correlations and the heritabilities of the two characters are known.
19. The improvement of performance in an environment if different from the one in which selection was carried out can therefore be predicted from a knowledge of the heritability of performance in each environment and the genetic correlation between the two performances.
20. In multiple trait selection appropriate weight-age is given to each character according to its relative economic importance, its heritability and the genetic and phenotypic correlations between the different characters.
21. Multiple trait selection including egg number and egg weight both.
22. 1.4 times.

I. B. Fill in the Blanks

1. Gene frequency
2. Fitness
3. (*a*) Intensity of selection; (*b*) Initial gene frequency
4. Reproduction
5. (*a*) Selection; (*b*) Breeding
6. Same
7. Response
8. Selection differential
9. Heritability and selection differential
10. Intensity of selection
11. Realized heritability
12. 15 to 30
13. Individual
14. Combined
15. Within family
16. Index selection
17. Restriction index
18. Additive
19. (*a*) General; (*b*) Specific combining ability
20. (*a*) Maternal; (*b*) Sex-linked.
21. Individual selection

II. A. True or False

1.	True	8.	False	15.	True	22.	True
2.	True	9.	True	16.	True	23.	True
3.	True	10.	True	17.	False	24.	False
4.	True	11.	True	18.	False	25.	True
5.	False	12.	True	19.	True	26.	False
6.	False	13.	True	20.	False	27.	False
7.	False	14.	False	21.	True	28.	True

II. B. True or False

1.	False	9.	False	16.	False	23.	True
2.	False	10.	False	17.	True	24.	True
3.	True	11.	True	18.	False	25.	True
4.	True	12.	True	19.	True	26.	False
5.	True	13.	True	20.	False	27.	True
6.	True	14.	True	21.	True	28.	True
7.	True	15.	True	22.	True	29.	False
8.	True						

Chapter 7

Poultry Breeding Systems

I. Fill in the Blanks

1. The coancestory of two individuals is the probability that two gametes taken at random one from each carry alleles that are identical by_______
2. The degree of relationship by decent between the two parents is called_______
3. The recurrence equation in self fertilization is_______
4. The recurrence equation in half-sib is_______
5. The coefficient of relationship between single first cousin is_______
6. The coefficient of relationship between single second cousin is_______
7. The constant value of rate of inbreeding in half-sibs is_______
8. Mating between identical genotype is called_______(Name the system of mating)
9. If the frequencies of dominant and recessive alleles are equal then the initial proportion of heterozygotes in population is_______

II. State True or False for Following Statements

1. The frequency distribution of most metric characters approximately more or less closely to normal curve.

2. The existence of additive genetic variance is a clear indication that the genes are acting additively.

3. Highly inbred lines or the F_1 of a cross between two such lines provide individuals all the identical genotype and therefore with no genetic variance.

4. When a character has high repeatability, multiple measurements give little gain in accuracy as there is large amount of special environmental variance.

5. In general, breeding for meat production is more complex than breeding for egg production.

6. Egg production and egg weight are generally positively correlated.

7. Many characters show more environmental variance among inbred than out bred individuals.

8. A character measured in two different environments is to be regarded not as one character but as two.

9. Existence of genotypic environmental interactions may mean that the best genotype in one environment is not the best in another environment.

10. The estimation of genetic correlations rests on the resemblance between relatives in a manner analogous to the estimation of heritability.

11. Intensity of egg production has a higher heritability than egg production up to 280 days age.

12. The standard deviation which measures the variability is a property of the character in a population and is expressed in standard deviation units.

13. Genetic correlations between traits are larger in magnitude than the corresponding phenotypic correlations.

14. The assumption that environmental variance is the same for all the genotypes, is not always true.

15. A correlation whatever its nature is the ratio of the appropriate covariance to the product of the two variances.

16. Dominance variance in poultry breeding is due to inter-allelic gene interaction.

17. For various broiler traits full exploitation of additive and non-additive genetic variance is not necessary.

18. In a poultry flock simultaneous improvement of four economic characters are desired. The best method of selection will be index method.

19. The condition under which inbreeding and crossing are likely to be better means of improvement than selection without inbreeding is when much of the genetic variance of the character is non-additive.

20. Layer type chickens are either inbred hybrids or strain crosses where as commercial broilers are usually breed crosses.

21. In correlation coefficient r (AB) if variance of A is equal to variance of B ($\sigma^{-2}A = \sigma^{-2}B$) then the r(AB) equivalent to the ordinary regression coefficient.

22. The correlation coefficient of an individual with itself is always one.

23. Heterosis is higher in back crosses than the three way and four way crosses.

24. Path coefficient has a physical unit.

25. In selfing, inbreeding coefficient increases with each generation.

26. Double first cousin are those whose parents belong to two sib-ships.

27. An individual can be homozygous without being autozygous.

28. The coancestory of an individual with unknown is equal or greater than one.

29. Inbreeding increases heterozygotes and decreases homozygous.

30. Inbreeding is often used because of necessity rather than choice, where stocks have reached on higher level of performance.

Answers

I. Fill in the Blanks

1. Descent
2. Coancestory/ Consnaguinity/ Coefficient of kinship
3. ½$[1+F_{t-1}]$
4. 1/8 $(1 + 6F_{t-1} + F_{t-2})$
5. 1/8
6. 1/32
7. 0.125
8. Selfing
9. 1/9

II. True or False

1.	True	9.	True	17.	False	25.	True
2.	False	10.	True	18.	True	26.	True
3.	True	11.	False	19.	True	27.	True
4.	False	12.	False	20.	True	28.	False
5.	True	13.	False	21.	True	29.	False
6.	False	14.	True	22.	True	30.	True
7.	True	15.	False	23.	False		
8.	True	16.	False	24.	False		

Chapter 8

Avian Molecular Genetics

I. A. Multiple Choice Questions (Tick Correct Answer/s)

Q. 1. Cellular DNA replication

(*a*) Requires the DNA double helix to be unwound

(*b*) Employs an enzyme called DNA ligase

(*c*) Occurs at a structure called a "growing fork"

(*d*) Often involves the synthesis of small pieces of RNA

(*e*) All of the above

Q. 2. Transfer RNA

(*a*) Is synthesized by a process known as translation

(*b*) Binds an amino acid covalently

(*c*) Contains a sequence of nucleotides known as a "codon"

(*d*) Is present only in eukaryotic cells

(*e*) None of the above

Q. 3. RNA

(*a*) Is more susceptible than DNA to degradation at high pH.

(*b*) Is found in all cells

(*c*) Has three primary roles in protein synthesis

(*d*) Contains subunits added one at a time during synthesis

(*e*) All of the above

Q. 4. The genetic code

(*a*) Is a triplet code

(*b*) Is degenerate

(*c*) Is "comma less"

(*d*) Was deciphered in the 1950s and 1960s

(*e*) All of the above

Q. 5. Which of the proteins are involved in translation?

(*a*) Topoisomerase and ribosomal RNA

(*b*) Initiation factors and topoisomerase

(*c*) Initiation factors, aminoacyl-tRNA synthetase and elongation factors

(*d*) All of the above

(*e*) None of the above

Q. 6. The wobble position of a codon

(*a*) Is the first (5′) base

(*b*) Is the third (3′) base

(*c*) Is the second base

(*d*) Often contain inosine

(*e*) None of the above

Q. 7. Operons

(*a*) Can be controlled by repressors

(*b*) Contain a DNA sequence called a promoter

(*c*) Contain genes

(*d*) All of the above

(*e*) None of the above

Q. 8. Primary transcripts of eukaryotic mRNA

(*a*) Are found in the cytoplasm

(*b*) Contains only introns

(*c*) Usually contains introns

(*d*) Are translated immediately

(*e*) None of the above

Q. 9. DNA synthesis begins

(*a*) At a single location in *E. coli*

(*b*) At a single location in the adenovirus genome

(*c*) At a single location in yeast

(*d*) At a site (s) that is G-C rich in *E. coli*

(*e*) All of the above

Q. 10. A protein that interacts with the origin of replication

(*a*) In SV40 is T antigen

(*b*) In *E. coli* is dnaA

(*c*) In adenovirus is terminal protein

(*d*) Unwinds DNA prior to replication

(*e*) All of the above

Q. 11. Form I DNA

(*a*) Is super-coiled

(*b*) Sediments faster than form II and III DNA

(*c*) Can be converted to form II by a nick in a single strand

(*d*) Can have more than one linking number

(*e*) All of the above

Q. 12. Topoisomerase I activity

(*a*) Cuts one strand of a DNA double helix

(*b*) Cuts both strands of a DNA double helix

(*c*) Does not change the linking number

(*d*) Changes the linking number by 2

(*e*) Requires energy supplied by ATP

Q. 13. Topoisomerase II activity

(*a*) Cuts both strands of a DNA double helix

(*b*) Changes the linking number by 2

(*c*) Requires energy supplied by ATP

(*d*) All of the above

(*e*) None of the above

Q. 14. The nucleotide sequences of replication origins

(*a*) Are conserved across bacteria

(*b*) Contains repeated elements

(*c*) Can be palindromes

(*d*) Can be functional when cloned into plasmid

(*e*) All of the above

Q. 15. Negative super-coiling in DNA

(*a*) Can be relieved by topoisomerase I nicking followed by re-ligation

(*b*) Is necessary for binding of *E. coli* RecA protein

(*c*) Results from under-winding in a circular molecule

(*d*) All of the above

(*e*) None of the above

Q. 16. DNA replication in eukaryotic cells

(*a*) Fills the same proportion of the generation time as it does in prokaryotic cells

(*b*) Is normally restricted to the S phase even in rapidly dividing cells

(*c*) Can be induced in G_2 by factors from S-phase cells

(*d*) Is not amenable to genetic analysis

(*e*) None of the above

Q. 17. Features in various models of recombination include

(*a*) Branch migration

(*b*) Single-strand nicks

(*c*) Double-strand breaks

(*d*) Synaptanemal Complexes

(*e*) All of the above

Q. 18. The technique known as fingerprinting can be used to compare samples of

(*a*) Amino acids (*b*) Lipid

(*c*) Protein (*d*) DNA

(*e*) None of the above

Q. 19. The most useful isotope in cellular auto radiographic studies is

(*a*) 14_C (*b*) 35_S

(*c*) 15_N (*d*) 32_P

(*e*) 3_H

Q. 20. Advances in which of the following fields have contributed greatly to the development and increased use of recombinant DNA methodology?

(*a*) Virology (*b*) Genetics

(*c*) Computer Science (*d*) Electrophoresis

(*e*) All of the above

I. B. Multiple Choice Questions (Tick Correct Answer/s)

Q. 1. Short segments of DNA, synthesized during DNA replication are called

(*a*) Okazaki fragments (*b*) DNA artifacts

(*c*) Primer dimers (*d*) Z DNA fragments

Q. 2. The process whereby DNA is used as template for the synthesis of RNA is called

(*a*) Translation (*b*) Transcription

(*c*) Replication (*d*) Reverse transcription

Q. 3. Both the strands of DNA double helix run_______to each other

(*a*) Horizontal (*b*) Vertical

(*c*) Parallel (*d*) Anti-parallel

Q. 4. DNA complementary to mRNA is called as

(*a*) z DNA (*b*) Junk DNA

(*c*) c DNA (*d*) r DNA

Q. 5. The most common DNA markers placed on avian genome are

(*a*) VNTRs (*b*) Microsatellites

(*c*) RFLPs (*d*) SNPs

Q. 6. The point at which replication is occurring is called

(*a*) Mutation point (*b*) Replication fork

(*c*) Transformation point (*d*) None of the above

Q. 7. The procedure for transferring denatured DNA from agarose gel to a nitrocellulose filter is called

(*a*) Northern blotting (*b*) Southern blotting

(*c*) Western blotting (*d*) Eastern blotting

Q. 8. RAPD technique was developed by

(*a*) Williams and co-workers (*b*) Welsh and McClead

(*c*) None of them (*d*) Both of the groups

Q. 9. The concentration of (A+G) in the DNA is

(*a*) is equals to the concentration of (T+C)

(*b*) is more than the concentration of (T+C)

(*c*) is less than the concentration of (T+C)

(*d*) have no relation with the concentration of (T+C)

Q. 10. VNTRs were reported by

(*a*) A. J. Jeffrey's and co-workers

(*b*) Y. Nakamura and co-workers

(*c*) Epplen and co-workers

(*d*) Epplen and Ali

Q. 11. The number of microsatellite markers mapped on chicken genome

(*a*) About 600 (*b*) About 500

(*c*) About 200 (*d*) About 300

Q. 12. Total number of base pairs of nucleotides on chicken genome are

(*a*) 1.2 billions (*c*) 2.2 billions

(*c*) 3.2 billion (*d*) 3.5 billions

II. Fill in the Blanks

1. Only the_______DNA enters the bacterial host cell after the infection.
2. The DNA mediated correction of genetic defects have been shown to occur in_______
3. The Watson and Crick model was based on_______of the nucleotides.
4. The chemical analysis showed that 1:1 ratio/ relationship do exists between_______and_______base.
5. The phase protein do not contain appreciable quantity of_______
6. The internal promoter may serve as transcription_______
7. Enhancers are simply called_______genes.
8. RNA polymerase is capable of polymerizing ribonucleoside triphophate into_______
9. Codon are essentially_______in nature
10. DNA dependent RNA polymerase is involved in_______
11. Three successive nucleotide base pair code for_______
12. The_______base of each codon is mostly responsible for the general character of the corresponding amino acid.

13. RFLP in animal breeding can serve as a reliable______

14. In______cases the cDNA is used for the hybridization in animals.

15. cDNA library is obtained by the______hybridization from cellular mRNA.

16. A coupling enzyme called______links an amino acid to the 3′ hydroxyl of the ribose of the terminal adenosine of a tRNA.

17. Short segments of DNA, synthesized during DNA replication, are called______.

18. ______is the name of the process whereby DNA is used as a template for the synthesis of RNA.

19. Enzymes that catalyze the synthesis of long, correctly ordered chains of subunits are generally classified as______.

20. The three generally recognized stages of the protein synthesis in both prokaryote and eukaryote are______and______.

21. A collection of bacteriophages containing inserted DNA sequences that re-representative of the entire genome of another organism is called______.

22. The technique known as______allows a researcher to enzymatically synthesize any piece of DNA that lies between the two known sequences.

23. DNA complementary to an mRNA sequence is called______.

24. A technique used to visualize the cellular location of specific mRNAs is called______.

25. Introduction of mutations, either deletions or point mutations, into a specific sequence of a cloned gene is called______

III. Write True or False Against Each Statement

1. A nucleotide triplet on a tRNA that forms a stable, hydrogen-bonding complex with a complementary triplet in mRNA is called codon.
2. A protein that assists in the binding mRNA to a ribosome is called an initiation factor.
3. The substrates needed for DNA synthesis are deoxyribonucleoside triphosphates and ribonucleoside triphosphates.
4. The strand of DNA that is synthesized as a continuous piece is the lagging strand; the strand that is synthesized in fragments is the leading strand.
5. The direction of DNA synthesis is $3' \rightarrow 5'$
6. During semi-conservative replication, a pre-existing strand is paired with a newly made strand.
7. A replication origin is a stretch of DNA necessary and sufficient to insure replication of a circular DNA in the appropriate host cell.
8. DNA synthesis occurs during the S phase of cell cycle in eukaryotes.
9. Nucleic acid segments containing RNA and DNA that are intermediates in DNA replication are called Okazaki fragments.
10. An enzyme called polymerase joins newly synthesized DNA fragments.
11. The RNA fragments that are necessary for the discontinuous synthesis of the lagging strand are produced by an enzyme called primase.
12. Replication of both paternal DNA strands occurs at the growing fork, which progress in the $3' \rightarrow 5'$ direction.
13. AUG is the initiation codon in mRNA.

14. The complex of all the proteins required for DNA synthesis is referred to as a(n) replisome.

15. A nonsense mutation generates a new UAG codon in the coding region of a gene.

16. The DNA region between two adjacent replication origins is called replicon.

17. The enzyme topoisomerase-I is required to separate newly replicated chromosomes.

18. The linked forms obtained after replication of circular genomes are called catenanes.

19. The ability of replication apparatus to correct in replication is called proofreading which is carried out by an exonuclease that works in the $3' \rightarrow 5'$ direction.

20. The enzyme primase unwinds the DNA double helix during replication.

21. The ability of the DNA polymerase to remain on the template and continue synthesis is called is its processivity.

22. The number of times one strand of a DNA double helix crosses the other in a molecule with fixed ends is the linking number.

23. Nucleic acid bases in tRNA are highly modified after synthesis; these modifications often involve the addition of a amino group to specific bases.

24. The parameter that describes the distance from a position where one strand of the DNA helix crosses the other to the next similar position is twist.

25. A cross-stranded intermediate in recombination is called a Holliday structure.

26. Some chain terminating mutations in bacteria are corrected if the gene containing the mutation is transferred to a suppressor strain.

27. Nonstrandard base-pairing between nucleotides in codon-anticodon pairs is called wobble.

28. The cellular structure that recognizes both tRNA and mRNA is called a ribosome.

29. During RNA processing in eukaryotes, sequences called exons are removed from the primary transcript.

30. One DNA strand serves as an templates, during DNA replication; short segment of RNA serve as primers during this process.

31. The sugar Moiety in RNA is Deoxyribose.

32. The genetic code is a triplet code

33. The utility of DNA variants in animal breeding is based on tight linkage between the marker and target gene.

34. Up take transforming DNA requires that its minimum molecular weight should be the order of 10°.

35. Single stranded DNA is much less easily incorporated in to the recipient cell.

36. Adenine and thymine are connected by 3 H bonds.

37. The genetic code places the amino acid adopter molecule in the reverse position during protein synthesis.

38. There was 3^4 possible triplet of codons for the stranded set of 20 amino acids.

39. Most of the amino acids can be specified by more than one codon.

40. Ambiguity of genetic codon may occur due to change in tRNA.

41. The poly-peptide chain is being synthesized from the free NH_2 end.

42. The presence of RFLPs allel could be correlated with the disease allel.

43. Mitochondrial region from where replication is believed to be initiated is termed L-Loop.

44. DNA library is a collection of dependently isolated vector linked DNA fragment derived from a single organism.

45. Chromosomal rearrangements generate variable number of tandom repeat called RSVs.

46. The natural variation in the DNA structure could easily be detected by the restriction fragment length polymorphism.

47. Codon cannot be misread due to environmental conditions around the tRNA and mRNA ribosome complex.

Answers

I. A. Multiple Choice Questions

1.	(*e*)	6.	(*b*)	11.	(*e*)	16.	(*b*)
2.	(*b*)	7.	(*d*)	12.	(*a*)	17.	(*e*)
3.	(*e*)	8.	(*c*)	13.	(*d*)	18.	(*d*)
4.	(*e*)	9.	(*a*)	14.	(*e*)	19.	(*d*)
5.	(*c*)	10.	(*e*)	15.	(*d*)	20.	(*e*)

I. B. Multiple Choice Questions

1.	(*a*)	4.	(*c*)	7.	(*b*)	10.	(*a*)
2.	(*b*)	5.	(*d*)	8.	(*d*)	11.	(*a*)
3.	(*d*)	6.	(*b*)	9.	(*a*)	12.	(*a*)

II. Fill in the Blanks

1. Phage
2. Eukarotype
3. Linear sequence
4. Purine and Pyrimidine
5. Phosphorous
6. Initiator
7. Modified zone
8. Polyribonucleotide
9. Universal

10. Genetic transcription
11. One amino acid
12. Middle
13. Marker
14. 90% cases
15. Reverse
16. Aminoacyl-tRNA synthetase
17. Okazaki fragments
18. Transcription
19. Polymerases
20. Initiation, Elongation and Termination
21. Library
22. Polymerase chain reaction
23. cDNA
24. In-situ hybridization
25. Site-directed mutagenesis.

III. State True or False

1	False	13	True	25	True	37	False
2	True	14	True	26	True	38	False
3	True	15	True	27	True	39	True
4	False	16	True	28	True	40	True
5	False	17	False	29	False	41	True
6	True	18	True	30	True.	42	True
7	True	19	True	31	False	43	False
8	True	20	False	32	True	44	False
9	True	21	True	33	True	45	False
10	False	22	True	34	False	46	True
11	True	23	False	35	True	47	False
12	False	24	True	36	False		

Chapter 9

General Avian Physiology

I. A. Multiple Choice Questions (Tick Out the Correct Option/s)

Q. 1. The cell membrane of the avian cell contains

(*a*) No protein but only fat

(*b*) Only proteins without any carbohydrate

(*c*) Lipoproteins with some amount of polysaccharides

(*d*) None of the above

Q. 2. DNA in an avian cell is present

(*a*) Only in the nucleus

(*b*) Only in the nucleolus

(*c*) In the nucleus mostly with some in mitochondria

(*d*) All of the above

Q. 3. Diffusion through the cell membrane varies according to

(*a*) Concentration gradient (*b*) M.W. of the substance

(*c*) Cross-section area (*d*) All of the above

Q. 4. One MOM is converted to osmotic pressure by multiplying with the factor

(*a*) 22 mmHg (*b*) 19.3 mmHg

(*c*) 100 mmHg (*d*) None of the above

Q. 5. Pinocytosis is necessary

(*a*) For transport of lipids through the avian cell membrane

(*b*) For transport of water through the cell membrane

(*c*) For transport of the Na through sodium channels

(*d*) None of the above

Q. 6. CO_2 is transported in the blood

(*a*) Mainly as HCO_3

(*b*) As carbomino-hemoglobin

(*c*) Combining with plasma protein

(*d*) All of the above

Q. 7. Avian respiratory system functions more efficiently because

(*a*) There are two diaphragms

(*b*) The diaphragms is of smooth muscle

(*c*) The diaphragms is not controlled by nervous system

(*d*) None of the above

Q. 8. The crop milk of the pigeon and dove is formed

(*a*) From the secretions of crop mammary glands

(*b*) From the exocrine glands of the crop mucosa

(*c*) From the disquamated epithelial cells of the crop mucosal epithelium

(*d*) None of the above

Q. 9. The ceca of the chicken are important because

(*a*) They are concerned with water absorption

(*b*) They digest all the proteins that comes from intestine

(*c*) They serve as a store for the undigested feed

(*d*) All of the above

Q. 10. The birds are Uricotolic because

(*a*) They excrete uric acid in the urine

(*b*) They consume a lot of uric acid and produce urea

(*c*) They can digest urea to produce microbial proteins

(*d*) None of the above

Q. 11. Optic lobes of the chicken are

(*a*) Much more developed compared to those of mammals

(*b*) Much less developed compared to those of mammals

(*c*) Absent in the brain

(*d*) None of the above

Q. 12. The respiratory function of the birds are dependent on and controlled by the

(*a*) Inflation and deflation of the lungs

(*b*) A vascular air sac present anteriorly and posteriorly

(*c*) Diaphragmatic movement during inhalation

(*d*) None of the above

Q. 13. Contraction of the glandular and muscular stomach of the fowl occur

(*a*) 2-3 per minute

(*b*) 10-16 per minute

(*c*) 12-18 per minute

(*d*) None of the above

Q. 14. Biphasic action potential in a stimulated avian nerve fibre can be seen

(*a*) When both the electrodes are placed on the surface of intact nerve fibre

(*b*) When both the electrodes are placed inside the nerve fibre

(*c*) When one electrode is inside and the other is outside the nerve fibre

(*d*) All of the above

Q. 15. Plasma osmotic pressure is mostly due to

(*a*) Plasma protein concentration

(*b*) Blood glucose concentration

(*c*) Plasma Na^+ concentration

(*d*) Blood urea concentration

Q. 16. The avian cardiac muscle cell membrane exchange

(*a*) Extracellular K^+ for intracellular protein

(*b*) Extracellular PO_4^+ with intracellular glucose

(*c*) Extracellular Na for intracellular Ca^{++} on an electrically neutral 1 to 2 basis

(*d*) All of the above

Q. 17. Summer heat stress causes the birds to lay less number of eggs due mostly to

(*a*) Poor feed intake

(*b*) Reduction in the function of reproductive-endocrine system

(*c*) More energy is spent for heat dissipation

(*d*) All of the above

Q. 18. Paravertebral sympathetic chain is present

(*a*) On either side of the spinal cord outside the CNS

(*b*) Embedded in the spinal cord

(*c*) Entirely unconnected with the spinal cord

(*d*) All of the above

Q. 19. There exists commonness either in origin or molecular structure of the hormone molecule of

(*a*) GH and Prolactin

(*b*) FSH, LH, & TSH

(*c*) ACTH, B-LPH & MSH

(*d*) All of the above

Q. 20. The hypothalamic neurohumour TRH causes

(*a*) The release of pituitary TSH in birds

(*b*) The release of pituitary prolactin in birds

(*c*) The gonadal growth in quails indicating gonado-trophin release

(*d*) All of the above

I. B. Multiple Choice Questions (Tick Out the Correct Option/s)

Q. 1. The mitochondrial matrix in addition to enzymatic components also contains

(*a*) Genetic material DNA

(*b*) Cellular machinery for protein synthesis

(*c*) Cellular reservoir for divalent Ca ions

(*d*) All above

Q. 2. Golgi apparatus is responsible for

(*a*) Synthesis of cell protein

(*b*) Synthesis of polysaccharide

(*c*) Synthesis of lipoprotein

(*d*) Synthesis of phospholipid

Q. 3. Microtubules form the core structure of

(*a*) Centrioles (*b*) Shafts of cilia

(*c*) Shafts of flagella (*d*) All above

Q. 4. The ceca in birds can be

(*a*) Paired (*b*) Single

(*c*) Rudimentary or absent (*d*) All above

Q. 5. The bile of chicken contains

(*a*) Proteases (*b*) Lipase

(*c*) Amylase (*d*) None above

Q. 6. The amino acid absorption in chicken depends on

(*a*) Rate of diffusion

(*b*) Presence of transporting electrolytes

(*c*) Cellular metabolism

(*d*) All above

Q. 7. The main difference in chemical constituents in urine in birds as compared to mammals is

(*a*) Preponderance of uric acid over urea

(*b*) Creatinine over creatin

(*c*) Creatinine over urea

(*d*) Urea over uric acid

Q. 8. The longest time spent by egg in female genital tract in chicken is in

(*a*) Magnum (*b*) Uterus

(*c*) Isthmus (*d*) Vagina

Q. 9. Fowl seminal plasma is almost completely lacking in

(*a*) Fructose and citrate

(*b*) Ergothioneine and inositol

(*c*) Phosphoryl choline and glyceryl phosphoryl choline

(*d*) All above

Q. 10. Secretion of HCL in the proventriculus is induced by

(*a*) Gastrin

(*b*) Cholocystokinin

(*c*) Both of the above

(*d*) None of the above

Q. 11. The birds excrete uric acid instead of urea because

(*a*) Urea is more toxic to tissues

(*b*) Urea is converted to uric acid in the kidney

(*c*) Urea is converted to uric acid in the liver

(*d*) Urea is converted to uric acid during storage in cloaca

Q. 12. The hormone related to water reabsorbtion in the kidney of birds is

(*a*) Angiotensin (*b*) Arginovasotosin

(*c*) Aldesterene (*d*) None of the above

Q. 13. The weight of testes in birds is about

(*a*) 0.5% of body weight (*b*) 1.0% of body weight

(*c*) 2.0% of body weight (*d*) 5.0% of body weight

Q. 14. The accessory sex glands in male birds are

(*a*) Cowper's gland (*b*) Seminal vesicles

(*c*) Both of the above (*d*) None of the above

Q. 15. The synthesis of nor-epinephrine is self limiting because

(*a*) Limited availability of phenylethanolamine-N-methyltransferase enzymes

(*b*) Nor-epinephrine inhibits the phenyl ethanol amine-N-methyltransferase

Q. 16. In birds the anterior pituitary secretes more prolactin because

(*a*) PRF from hypothalamus

(*b*) PIF and PRF from hypothalamus

Q. 17. The most appropriate bioassay of prolactin is

(*a*) Broodiness in birds

(*b*) Pigeon crop milk secretion

Q. 18. Progesterone in the birds is secreted from

(*a*) Corpus luteum (*b*) Granulosa cells

Q. 19. Formation of HCO_3 in the shell glands is catalyzed by

(*a*) Alkaline phosphatase (*b*) Carbonic anhydrase

Q. 20. In female chicken

(*a*) Both the ovaries and oviducts are functional

(*b*) Only left ovary and left oviduct are functional

(*c*) Only right ovary and its oviduct are functional

(*d*) All the above

Q. 21. Oviposition refers to

(*a*) Liberation of ovum by rupture of Graffian follicle

(*b*) Laying of egg by the bird

(*c*) Formation of egg in the uterus

(*d*) Passage of egg in the oviduct

Q. 22. Parthenogenesis which refers to embryo development in a unfertilized oven is common in

(*a*) Chicken (*b*) Duck

(*c*) Guinea fowl (*d*) Turkey

Q. 23. Sterile mule ducks are the product of

(*a*) Cross of white Pekin with Aylesbury

(*b*) Cross of Rouen with Cayuga

(*c*) Muscovy with other breeds

(*d*) White Pekin with other breeds

Q. 24. The true stomach of chicken is

(*a*) Crop (*b*) Gizzard

(*c*) Pro-Ventriculus (*d*) None of the above

Q. 25. Bile salts are essential

(*a*) For breakdown of polysaccharides in to Mono-saccharide

(*b*) For breakdown of proteins into amino-acids

(*c*) For emulsification of fats

(*d*) None of the above

Q. 26. The voice of fowl is produced in

(*a*) Upper larynx (*b*) Lower larynx

(*c*) Trachea (*d*) Air sacs

Q. 27. The name of the only gland present in fowl is

(*a*) Uropygeal gland (*b*) Oil gland

(*c*) Preen gland (*d*) All of above

Q. 28. During normal egg formation, the time taken by egg in magnum is

(*a*) 1 hr (*b*) 2 hr

(*c*) 3 hr (*d*) 4 hr

II. Fill in the Blanks

1. Somatostain inhibits the secretion of _______and_______normally.
2. The enzyme .... catalyses the conversion of nor-epinephrine to epinephrine in the adrenal medulla.
3. Exchange of gasses in avian lung results from_______and is governed by physical laws.
4. Respiratory centre in the birds is located in the_______of the avian brain.
5. Lesion in the_______nucleus of avian hypothalamus causes hyperphagia
6. The female birds of the same breed show_______(higher/ lower) body temperature than the males under similar environmental conditions.
7. HCL is secreted by_______cells of proventriculus in birds.
8. Activity of trypsin in birds is optimum at a pH of_______
9. Bile is transported from the gall bladder to the duodenum through_______
10. The site of digestion of crude fibre in birds is the_______
11. The source of thromboplastin is_______
12. The life span of fowl erythrocytes is about_______
13. The heart rate of the day old chick is about_______
14. Resorption of useful components of urine takes place at_______

15. The role of ICSH is_______
16. The semen of cock is usually_______(acidic/ alkaline)
17. The egg usually remains in the uterus of birds for_______
18. The role of prolactin in birds is_______
19. Thyrotrophic hormone is produced by_______
20. The cells of pancreas are concerned with_______
21. Significant amount of starch digestion in chicken occurs in_______
22. Sham feeding of chicken results in an increased_______
23. Avian secretin has been shown to stimulate_______
24. The most important site for absorption of end products of digested feed in birds is_______
25. Normal rhythmic respiration is maintained by_______
26. Section of both vagi in birds_______respiratory frequency.
27. Haldane effect indicates what_______
28. Chemoreceptors in the carotid bodies of birds respond to_______
29. The most prominent pneumatic bone in domestic birds is_______
30. Birds increase evaporation by a mechanism named 'gular flutter' which means_______
31. Behavioural thermo-regulatory responses in birds are regulated by_______region of the brain.
32. Cardiac output is defined as_______
33. The ECG indicates_______
34. A rectal temperature of_______°C appears to be upper limit of safety in the hen.

35. Reproduction in birds is characterized by ovoviviparity which means_______

36. Length of ovulatory cycle of domestic hen is_______

37. The autonomic nervous system includes those efferent nerves that innervate_______

38. Parasympathetic fibres leave in the cranial nerves and sacral nerves and synapse on_______

39. The sequence of events occurring in a complete cardiac cycle includes_______and_______.

40. Acclimatization refers to_______

41. In a cold environment, the chicken reduces its heat loss by "hunching" which means_______

42. Synapse is defined as_______

43. White matter in the central nervous system is formed by_______

44. Both inspiration and expiration are active processes in the birds and depend on_______

45. Birds are also known as 'endotherms' which indicates_______

46. At higher temperature most heat is lost by_______

47. Oviposition is controlled by_______

48. The egg albumin is formed in the_______of oviduct.

49. The colours which cause maximum release of LH are_______

50. Premature ovulation will be there if progesterone is injected in_______area of hypothalamus.

51. The total number of air sacs in birds are_______

52. The oxygen dissociation curve of birds is towards _______of mammalian.
53. Panting centre in the birds is located in_______
54. The range of thermo-neutral zone in birds is_______
55. The ceca are well developed in_______eating birds.
56. Chalaziferous layers are formed due to_______.
57. In animals millions of cells are organized into specialized_______that interact to sustain life.
58. Biological systems are broadly divided into eukaryotic cells_______and viruses.
59. The control of the protein synthesis by_______is mediated by action of RNA.
60. The crop milk secretion in pigeons and dove is dependent on_______.
61. In birds trypsin and chymotrypsin are secreted from_______.
62. The amount of water consumed_______significantly in laying hens as compared to non-laying period.
63. The shorter life span of avian erythrocytes compared to mammalian is related to_______avian body temperature.
64. Most birds have two main coronary arteries but some may have_______.
65. Plumage of birds provides them with effective_______
66. Natural day light or artificial light initiates the production of semen in males by stimulating the release of_______
67. Insertion of a glass catheter in the shell gland/uterus of adult chicken results in_______
68. The postovulatory follicles in chickens do not form_______

III. State if the Following Statements are True or False

1. The thickness of the left ventricle wall is more than that of right ventricle even if the intraventricular pressure in both is same.
2. Avian hypothalamic GnRH causes selectively the release of hypophysial FSH.
3. Troponin C is a component protein of myosine (Actin) to which Ca++ is bound.
4. Muscarinic effect is the same as the nicotinic effect of acetylcholine.
5. Progesterone is a polypeptide hormone that is secreted from theca cells of chicken ovary.
6. Estrogen facilitates Ca++ mobilisation from the bone in a laying hen.
7. Carbonic anydrase has enough roles to play in the process of egg shell formation.
8. Chickens possess a Na^{+} dependent mobile carrier system for active transport of amino acids similar to that of mammals.
9. Theca interna cells of ovarian follicles have many LH receptors and LH acts via cyclic AMP to convert cholesterol to androsteneione.
10. The amino acid content of oxytocin and avian vasotocin are almost same having equal physiological action.
11. Gastric acid secretion in birds is stimulated by histamine via H_2 receptors and by actyl choline via muscarinic receptors.
12. The avian spermatozoa are cold–shock resistant.
13. All the birds have equally developed crop that secrets many digestive enzymes.

14. Capacitation of avian spermatozoa is a must before the sperms can fertilize the ovum.

15. Melatonin is secreted from adenohypophysis of the turkey and influences the maturity of the females.

16. High temperature with low humidity is better than low temperature with high humidity because the bird can not sweat in the latter condition.

17. Photoreceptors are present in the avian brain outside the retina.

18. Participation of kidney is important for the action of Vitamin D_3 in chicken.

19. Phospholipids such as lecithin and cephalin are found in substantial quantities in egg yolk.

20. The body temperature of hatched chicks is higher than the adults as they have smaller body mass.

21. The pH of the contents in gizzard ranges from 4 to 6.

22. The pancreatic juice of birds contains invertase and lipase.

23. Female constringes have the same number of RBC as normal females.

24. The main site of erythrocyte destruction in birds is the spleen.

25. The vagus has stimulatory effect on heart contraction.

26. The two ureters of birds open into the bladder.

27. In birds the cortex and medulla of kidney are not well differentiated.

28. The glomerular filtrate contains only the excretory products which the bird does not require.

29. The sperms of cock undergo maturation changes during their storage in epididymis.

30. The release of LH from anterior pituitary is under the control of hypothalamus.

31. The pigment of the shell is formed in the uterus.

32. After ovulation the egg is fertilized usually within 2-3 hours.

33. Egg formation requires 18-20 hours.

34. The plumage of birds provides an effective barrier to heat loss from the skin surface to the surroundings.

35. Thermal polypnea decreases tidal volume.

36. Moulting is associated with an increase in activity of adrenal gland.

37. The birds lack Enterokinase enzyme.

38. Gall bladder is not present in all avian species.

39. Progesterone in birds is produced by the ovary.

40. The blood of laying hen contains about 12-15 mg calcium.

41. Propagation of action potential is faster in myelinated fibre.

42. The lungs of birds do no expand and contract during the respiratory cycle.

43. Air sacs in the birds function in gas exchange.

44. *Pars intermedia* is present in the birds.

45. MSH is produced from posterior pituitary of birds.

46. Hypophysectomy in birds increases the reproductive performance.

47. In birds optic tectum is not very well developed.

48. Aldosterone helps in increased excretion of sodium and water.

49. Corticosterone treatment inhibits the growth of bursa of Fabricious.

50. The Blood-Brain-Barrier is very well developed in newly hatched chicks.

51. Nasal glands are located in the nasal cavity of birds.

52. Pons is absent in the birds.

53. The average lifespan of the avian RBCs is more than 100 days.

54. Nucleus is not the only repository of genetic information in the eukaryotic cell.

55. Actin containing microfilaments of muscle cells have been found associated with myosin, tropomyosin and alpha-actinin.

56. Glandular portion of stomach in birds is Gizzard.

57. The removal of gizzard has profound effect on digestion even if the feed provided is soft.

58. Most of the lipids present in birds eggs are synthesized from non-lipid constituents.

59. The method of TLC counting of avian blood is just like mammalian blood.

60. Avian lungs contract and relax like mammalian lungs.

61. Removal of the ruptured follicle causes delayed oviposition of the ovum released from it.

62. Progesterone is the primary steroid secreted from largest graffian follicle.

63. A shell membrane is laid down around albumin as it passes through isthmus.

64. LH surges occur during the dark period for induction of ovulation.

65. Accessory sex glands in male birds are absent.

66. Testes in birds have no role in the regulation of FSH secretion from adenohypophysis.

67. Sperms are stored in uterovaginal gland in birds.

68. AI should be carried out every 7 to 14 days to get fertilized egg.

IV. Match the Following Pairs

1.	Exteroceptive reflex	(*a*)	Part of PRL (Prolactin?) releasing mechanism in birds
2.	Factor XI	(*b*)	Placing birds in uncomfortable surrounding
3.	Serotonin	(*c*)	No role in sperm maturation
4.	Livetins	(*d*)	Water soluble egg proteins.
5.	Secretion of albumen	(*e*)	Plasma Thrombo-plastin antecedent (PTA)
6.	Discouraging broodiness	(*f*)	Narrow and long
7.	Juxta testicular structures	(*g*)	Nasal gland
8.	Sperm head	(*h*)	Cutaneous receptor stimulation
9.	Extra-renal salt secretion	(*i*)	Function of magnum
10.	Inulin	(*j*)	GFR (Glomerular Filteration Rate)

Answers

I. A. Multiple Choice Questions

1	(*c*)	6	(*d*)	11	(*a*)	16	(*c*)
2	(*c*)	7	(*d*)	12	(*b*)	17	(*d*)
3	(*d*)	8	(*c*)	13	(*a*)	18	(*a*)
4	(*b*)	9	(*a*)	14	(*a*)	19	(*d*)
5	(*d*)	10	(*a*)	15	(*a*)	20	(*d*)

I. B. Multiple Choice Questions

1	(*d*)	8	(*b*)	15	(*b*)	22	(*d*)
2	(*a*)	9	(*d*)	16	(*a*)	23	(*c*)
3	(*a*)	10	(*c*)	17	(*b*)	24	(*c*)
4	(*a*)	11	(*a*)	18	(*b*)	25	(*c*)
5	(*d*)	12	(*b*)	19	(*b*)	26	(*b*)
6	(*a*)	13	(*b*)	20	(*b*)	27	(*d*)
7	(*a*)	14	(*d*)	21	(*b*)	28	(*c*)

II. Fill in the Blanks

1. GH, TSH
2. PNMT
3. Diffusion
4. Medulla
5. Ventromedial

6. higher
7. Chief cells
8. 6-7
9. Cystic duct
10. Caeca
11. Tissue juices
12. 28-35 days
13. 280-300 per minute
14. Kidney tubules
15. Growth and development of leydig cells
16. Alkaline
17. 18-20 hrs
18. Suppression of FSH and inducing broodiness in hens
19. Large beta cells
20. Formation of glycogen
21. Crop
22. Gastric secretion
23. Pancreatic secretion
24. Upper ileum
25. Afferent fibres of vagi
26. Decreases
27. The degree of oxygenation of Hb greatly influences the quantity of CO_2 that the blood contains.
28. Low O_2 pressure.
29. Humerus.
30. Rapid oscillation of the thin floor of the mouth and the upper part of the throat.
31. Preoptic region of brain.
32. The amount of blood ejected from the ventricles in a given time.

33. Electrical activity (depolarization and repolarisation) of the heart picked up from electrodes attached to parts of the body other than the heart itself.
34. 45 °C
35. That the potential offspring leaves the mother's body after a very short period of embryonic development and undergoes the major portion of its development outside her body.
36. 25-26 hrs.
37. Structures other than the skeletal muscle.
38. Scattered small ganglion on the innervated organ
39. Systole & Diastole
40. Seasonally induced changes resulting form climatic variation usually out of doors.
41. Reduces its surface area.
42. Site of integrative function of the nervous system.
43. Myelinated axons and neuroglia form white matter. Gray matter is formed by un-myelinated axons with dendrites and perikarya.
44. Contraction of respiratory muscles.
45. They are able to increase their body temperature by generation a considerable amount of heat within their tissues.
46. Evaporative cooling.
47. Arginine vasotocin
48. Magnum
49. Red and orange
50. Pre-optic
51. Nine
52. Left
53. Medulla oblongata
54. 25-37

55. Grain
56. Mechanical twisting
57. Tissue
58. Prokaryotic cells
59. Ribosomes
60. Prolactin hormone
61. Pancreas
62. Increases
63. Higher
64. Three
65. Insulation
66. FSH and LH
67. Deposition of $CaCO_3$ crystals
68. Corpus luteum

III. True or False

1.	False	13.	False	25.	False	37.	True
2.	False	14.	False	26.	False	38.	True
3.	True	15.	False	27.	True	39.	True
4.	False	16.	False	28.	False	40.	False
5.	False	17.	True	29.	False	41.	True
6.	True	18.	True	30.	True	42.	True
7.	True	19.	True	31.	True	43.	False
8.	False	20.	False	32.	False	44.	False
9.	True	21.	False	33.	False	45.	False
10.	False	22.	True	34.	True	46.	False
11.	True	23.	True	35.	True	47.	False
12.	True	24.	False	36.	False	48.	False

49.	True	54.	True	59.	False	64.	True
50.	False	55.	True	60.	False	65.	True
51.	False	56.	False	61.	True	66.	False
52.	True	57.	False	62.	True	67.	True
53.	False	58.	True	63.	False	68.	False

IV. Match the Following

1.	(*h*)	4.	(*d*)	7.	(*c*)	10.	(*j*)
2.	(*e*)	5.	(*i*)	8.	(*f*)		
3.	(*a*)	6.	(*b*)	9.	(*g*)		

Chapter 10

Avian Stress and Environmental Physiology

I. A. Multiple Choice Questions (Tick Out the Correct Statement/s)

Q. 1. In a cold environment the chicken reduces its heat loss

(*a*) By hunching

(*b*) By tucking the head under wings

(*c*) By sitting

(*d*) All the above

Q. 2. Heat production is improved in birds by

(*a*) Increasing the temperature of thermo-receptors of C.N.S.

(*b*) By cooling the thermo-receptors of C.N.S

(*c*) Bilateral lesions of anterior hypothalamic pre-optic region.

(*d*) None of the above.

Q. 3. At high air temperatures in flight most heat is lost by

(*a*) Evaporative cooling

(*b*) Increase in respiratory volume

(*c*) Increased surface area

(*d*) All the above

Q. 4. When birds become active there is an immediate

(*a*) Dissipation of more heat

(*b*) Conservation of heat

(*c*) Rise in body temperature

(*d*) None of the above.

Q. 5. The insulating value per unit thickness of feather is

(*a*) Poor or low (*b*) High

(*c*) Not satisfactory (*d*) Insignificant

Q. 6. The heart rate at a very high body temperature

(*a*) Remains unaffected (*b*) Not much affected

(*c*) Rises to a high value (*d*) Becomes slow

Q. 7. Birds that live in a cold climate have

(*a*) Larger body size (*b*) Smaller body size

(*c*) No effect on body size (*d*) None of the above

Q. 8. On exposure to heat fowls become

(*a*) Exceedingly agitated

(*b*) Exceedingly sluggish

(*c*) Poorly agitated

(*d*) Hypothermic

Q. 9. Birds fly enormous distance to

(*a*) To ensure food supply

(*b*) To avoid unfavourable environment

(*c*) To maintain homeothermy

(*d*) All the above

Q. 10. An insulated floor in pen

(*a*) Has no effect on thermoregulation

(*b*) Temperature of the ground floor affects more than that of air.

(*c*) Temperature of the air affects more than that of ground floor.

(*d*) None of the above.

Q. 11. In order to maintain a relatively constant temperature of body, heat must be

(*a*) Lost (*b*) Conserved

(*c*) Both (*d*) None of the above

Q. 12. The 47 °C deep body temperature of chickens is the temperature of

(*a*) Upper lethal (*b*) Thermo neutral

(*c*) Lower lethal (*d*) None of the above

Q. 13. An increase in the body temperature of the adult birds than the chicks is due to

(*a*) Feather coverage

(*b*) Increased metabolic activity

(*c*) Size

(*d*) All the above

Q. 14. The movement of feathers is produced by

(*a*) Voluntary muscle

(*b*) Involuntary muscle

(*c*) Both voluntary and involuntary muscles

(*d*) None of the above

Q. 15. The air sacs in poultry are

(*a*) Largely non vascular

(*b*) Largely vascular

(*c*) Both vascular and non vascular

(*d*) None of the above

Q. 16. The feature of acclimatization in birds is

(*a*) A reduction in haematocrit and plasma value

(*b*) An increase in haematocrit and plasma value

(*c*) A reduction in haematocrit and increase in plasma volume

(*d*) None of the above

Q. 17. In the birds, native to high altitude, specific haemoglobin O_2 affinity

(*a*) Increases (*b*) Decreases

(*c*) None of the above

Q. 18. The chickens hold their wings a little apart from the body

(*a*) Under cold stress (*b*) Under heat stress

(*c*) None of the above

19. **The cold stress brings changes in body**

(*a*) Neural only (*b*) Humoral only

(*c*) Neurohumoral (*d*) None of the above

Q. 20. The sensible heat loss from the body is

(*a*) By radiation and convection

(*b*) By convection and conduction

(*c*) By radiation, convection and conduction

(*d*) None of the above

I. B. Multiple Choice Questions (Tick Out the Correct Statement/s)

Q. 1. A continuous exposure of day light for 24 hours to domesticated fowl may results

(*a*) No interruption in egg production

(*b*) Interruption in egg production

(*c*) None of the above

Q. 2. An abrupt change in the environmental temperature influences

(*a*) Egg weight

(*b*) Egg weight and shell quality

(*c*) Egg weight, shell quality and egg production

(*d*) None of the above

Q. 3. The thyroid gland activity is increased in the climatic condition of

(*a*) Winter
(*b*) Summer
(*c*) Rainy season
(*d*) None of the above

Q. 4. To overcome stress of high environmental temperature, the body activities are

(*a*) Increased
(*b*) Decreased
(*c*) None of the above

Q. 5. The large scale climatic conditions of the open atmosphere in an area is termed to

(*a*) Microclimatic
(*b*) Macroclimatic
(*c*) Weather
(*d*) None of the above

Q. 6. At high altitude, the environment has

(*a*) Low atmospheric and oxygen pressure

(*b*) High atmospheric and low oxygen pressure

(*c*) Low atmospheric and high oxygen pressure

(*d*) None of the above

Q. 7. The small sized avian breeds are found in

(*a*) Cooler climate
(*b*) Warmer climate
(*c*) None of the above

Q. 8. The melanin in the skin absorbs erythematous radiation

(*a*) Partially
(*b*) Completely
(*c*) None of the above

Q. 9. The hours of day light or artificial light in bird

(*a*) Induces ovulation (*b*) Inhibits ovulation

(*c*) None of the above

Q. 10. An increased amount and hours of day light or artificial light in hens

(*a*) Increases egg production

(*b*) Decreases egg production

(*c*) None of the above

II. Fill in the Blanks

1. The core temperature of the bird is measured by inserting a thermocouple probe in to______
2. Fluctuation in body temperature during the course of day is known as______
3. The process of changes induced by prolonged exposure to a particular temperature is known as______
4. Peripheral tissue freezing in low enough environmental temperature is circumvented by a phenomenon called______
5. Heat lost from the surface of body is both evaporative and ______means.
6. Down feathers of plumage provide effective barrier______
7. Most conspicuous thermoregulatory behaviour of birds is migration to______
8. Mammals use burrows and shelters against cold but birds usually______
9. The regulation of body temperature involves thermo-sensitive neurons of CNS and peripheral______
10. In birds peripheral warm receptors are present in the______

11. Sweat glands are_______in birds.
12. Chicks of domestic chickens are covered with _______feathers and are able to respond efficiently and effectively to cold and heat.
13. Body temperature of birds varies between 104 °F to_______°F.
14. Rate of respiration depends upon central_______and not on reflexes from skin.
15. The thermal insulation in penguin is owing to_______
16. Feather_______has a considerable effect on cold tolerance.
17. The temperature difference between skin and surface of feather is much_______than between rectum and skin.
18. Wings of birds are_______in cool climate.
19. Melanin in feather pigmentation_______in warm and humid climate.
20. The birds become restless at a_______temperature to around 111 °F.
21. Birds are derived from pikilothermic reptiles, but they are_______in their body temperature control.
22. The oil producing_______gland is located over the last vertebra of the birds.
23. The air passages in birds are_______at both ends, which permits free air circulation.
24. The extensive surface areas of air sacs in birds permit_______cooling.
25. The high ambient_______depresses quality of the egg shell.
26. The more effective wave length of colour light is_______
27. A change in incubation temperature in the early cleavage stage of chick embryo produces_______twinning.

III. Write if the Following Statements True or False

1. Birds are homeotherms and regulate their body temperature by relying on heat gained from surroundings.
2. Salt glands of some of the birds enable them to avoid consequences of dehydration.
3. The temperature regulation in birds can be discussed by equation $M = E \pm R \pm C \pm K \pm S$.
4. The circadian rhythm of deep body temperature fluctuates according to size of the bird.
5. The daily cycle of body temperature has no relationship to photoperiodicity.
6. Acclimatization refers to changes induced by prolonged exposure to a particular temperature in laboratory conditions.
7. The body temperature of dehydrated ostrich increases during exposure to heat in contrast to hydrated bird.
8. In birds and mammals heat production per unit body weight is less in larger breeds.
9. Air temperature higher than the upper critical temperature in normal birds decreases oxygen consumption to bring about thermoregulation.
10. Counter-current heat exchange usually occurs in the limb in the cold environment.
11. The deposition of fat toward off cold climate in pigeons acclimated to environmental temperature of 10 °C is greater than pigeons acclimated to 20 °C.
12. The insulation provided by plumage is lesser in larger birds than in small birds.
13. Usually thermal conductance of small birds is greater, thus insulation is less than that of large birds.

14. The extensive air sac system of birds helps in respiration from lung tissue and evaporative heat loss.
15. *Gular flutter* in some birds is an effective heat loss mechanism as it circumvents hyper-ventilation.
16. Peripheral temperature receptors and associated neurons have no role in temperature regulation.
17. Thermal polypnea occurs when deep body temperature of fowl increases.
18. Some of the heat produced during flight is stored in the bird.
19. The respiratory minute volume during flight is five times than when the bird is rest.
20. Newly hatched chicks have higher body temperature than adult birds.
21. The internal body temperature of poultry is lower than that of mammals.
22. White leghorn males have higher body temperatures than females in cooler environments.
23. The skin of birds does have sebaceous glands.
24. Uropygeals of water fowl are larger than those of chickens.
25. At panting in birds, tidal volume increases.
26. Hypothermia results in a fall of blood pressure.
27. The percentage of O_2 is more in the abdominal air sacs than the inter-clavicular air sacs in chickens.
28. Heat loss in birds is more from their dorsal surface.
29. In hot days, birds cool their eggs by standing over them without contact.
30. The brood patches in birds facilitate the transfer of heat.

31. An increase in the hours of day length to pullets induces early sexual maturity.

32. The physical environment significantly affects total number of egg production per clutch.

33. The seasonal variation affects equally yolk and albumen components of the egg.

34. An exposure to intense solar radiation is an internal stress.

35. White or cream coloured hair largely absorbs infrared wave length.

36. The bills are relatively shorter in warmer climate than that in cooler climate.

37. Hens generally lay their eggs in the hours of day light.

38. The respiratory diseases, as internal stress factor decrease the viscosity of albumen.

39. A deficiency of calcium and vitamin–D results into thinning of egg shell.

40. Medullary bone does exist in male birds.

41. Hours of day light induces production of semen in birds.

42. The release of FSH from the anterior pituitary does not influenced by the photoperiodism.

43. The decline in fertility due to high temperature in summer is generally due to decrease in sperm concentration in birds.

44. Shorter wave length light is less effective in influencing testes of birds as compared to longer wave length.

Answers

I. A. Multiple Choice Questions

1	(*d*)	6	(*c*)	11	(*c*)	16	(*a*)
2	(*b*)	7	(*a*)	12	(*a*)	17	(*a*)
3	(*d*)	8	(*b*)	13	(*d*)	18	(*b*)
4	(*a*)	9	(*d*)	14	(*a*)	19	(*c*)
5	(*b*)	10	(*b*)	15	(*a*)	20	(*c*)

I. B. Multiple Choice Questions

1	(*b*)	4	(*b*)	7	(*b*)	10	(*a*)
2	(*c*)	5	(*b*)	8	(*b*)		
3	(*a*)	6	(*a*)	9	(*a*)		

II. Fill in the Blanks

1. Rectum or stomach
2. Circadian rhythm
3. Acclimation
4. Cold vasodilatation
5. Non-evaporative means
6. To heat loss from skin
7. Warmer or cooler areas
8. Do not
9. Temperature receptors

10. Beak
11. Absent
12. Down or down feathers
13. 109
14. Temperature
15. Subcutaneous fat or blubber
16. Erection
17. Greater
18. Longer
19. Increases
20. Rectal
21. Homeothermic
22. Uropygeal or preen
23. Open
24. Evaporative
25. Temperature
26. Red
27. Increased

III. State True or False

1	False	12	False	23	False	34	False
2	True	13	True	24	True	35	False
3	True	14	True	25	False	36	True
4	True	15	True	26	True	37	True
5	False	16	False	27	True	38	True
6	False	17	True	28	False	39	True
7	True	18	True	29	True	40	False
8	True	19	False	30	True	41	True
9	False	20	False	31	True	42	False
10	True	21	False	32	True	43	True
11	True	22	True	33	False	44	True

Chapter 11

Incubation and Hatching

I. A. Multiple Choice Questions (Tick Correct Answer/s)

Q. 1. The requirement of relative humidity in the egg holding room is

(*a*) 75-80% (*b*) 70-75%

(*c*) 60-70% (*d*) 80-95%

Q. 2. Most popular method of sexing the chicks at day old stage is:

(*a*) Feather sexing (*b*) Colour sexing

(*c*) Vent sexing (*d*) Machine sexing

Q. 3. The most important egg borne disease transmitted through hatching eggs from breeders to chicks is:

(*a*) Aspergillosis

(*b*) Mycloplasmosis

(*c*) Infectious bronchitis

(*d*) Marek's diseases

Q. 4. The best disinfectant used for the disinfection of hatchery is:

(*a*) Quaternary ammonium compounds

(*b*) Phenolic compounds

(*c*) Formaldehyde

(*d*) Hypochlorites

Q. 5. During the embryonic development 1st meiotic division takes place:

(*a*) Prior to ovulation of the follicle

(*b*) After the ovulation of the follicle

(*c*) At the time of ovulation of the follicle

(*d*) After the fertilization of ovum

Q. 6. For good hatchability eggs selected should be

(*a*) Above 60 g (*b*) 45 to 50 g

(*c*) 50 to 55 g (*d*) 55 to 60 g

Q. 7. For obtaining highest percentage of fertile eggs in broilers, there should be male to female ratio of:

(*a*) 1:10 (*b*) 1:5

(*c*) 1:12 (*d*) 1:15

Q. 8. For the best hatching results the storage period of eggs should be:

(*a*) As shorter as possible (*b*) 3 days

(*c*) One week (*d*) 10 days

Q. 9. For obtaining better hatchability the relative humidity in hatcher should be:

(*a*) 90% (*b*) 80%

(*c*) 60-70% (*d*) 50-60%

Q. 10. Artificial incubation was first started in

(*b*) Egypt and China (*c*) Japan

(*d*) USA (*e*) England

Q. 11. The concentration of CO_2 in the incubator for best results should be:

(*a*) 0.3-0.5% (*b*) 0.5-1.0%

(*c*) 2.0-3.0% (*d*) 3.0-5.0%

Q. 12. For best results the turning of eggs is essential and egg trays should be tilted to:

(*a*) At 60° angle

(*b*) At 45° angle

(*c*) At 30° angle

(*d*) At 65° angle

Q. 13. The incubated eggs should be candled on:

(*b*) 16th or 18th day (*c*) 8th or 9th day

(*d*) 5th or 7th day (*e*) 3rd or 4th day

Q. 14. At high altitude the poor hatchability may be because of:

(*b*) Low oxygen pressure

(*c*) High carbon dioxide

(*d*) High nitrogen pressure

(*e*) A combination of above two (*b* and *c*)

Q. 15. For high fertility the age of the breeder males should be:

(*a*) 10 weeks (*b*) 6 or 7 months

(*c*) One year (*d*) 10 months

Q. 16. Hatchability is high in:

(*a*) High egg producers

(*b*) Low egg producers

(*c*) Moderate egg producers

(*d*) Old birds

Q. 17. The temperature in the incubator should be for best results:

(*a*) 99.5°F (*b*) 98.4°F

(*c*) 37°F (*d*) 90°F

Q. 18. Along with the age of breeder stock, the most important factors for good fertility is:

(*a*) Temperature of the house

(*b*) Relative humidity in the house

(*c*) Nutrition of the bird

(*d*) Kind of the breeds

Q. 19. Restricted feeding in broiler breeder is essential for high fertility from:

(*a*) 8 to 20 weeks of age

(*b*) 10 to 30 weeks of age

(*c*) 8 to 30 weeks of age

(*d*) 8 to 40 weeks of age

I. B. Multiple Choice Questions
(Tick Correct Answer/s)

Q. 1. Correct temperature in a forced draft incubator should be

(*a*) 99.5°F (*b*) 103°F

(*c*) 90°F (*d*) None of the above

Q. 2. Correct wet bulb reading on the correct dry bulb reading should be

(*a*) 85°F (*b*) 60° F

(*c*) 90°F (*d*) None of the above

Q. 3. How many times hatching eggs should be turned in 24 hours in a setter

(*a*) Two times (*b*) Six times

(*c*) Twelve times (*d*) Ten times

Q. 4. Hatching eggs should not be stored for more than following days for good results

(*a*) 7 days (*b*) 5 days

(*c*) 20 days (*d*) 10 days

Q. 5. Humidity is maintained more in

(*a*) Setter (*b*) Hatcher
(*c*) Shed (*d*) Slaughter house

Q. 6. Chicks come out on 21st day

(*a*) In Natural hatching
(*b*) In forced draft incubators
(*c*) Still air incubators
(*d*) In all

Q. 7. How many eggs can be hatched by a broody hen at a time?

(*a*) 11 (*b*) 21
(*c*) 8 (*d*) 5

Q. 8. Chicks hatch by making holes in the shell by a false tooth in

(*a*) Circular way (*b*) Irregular way
(*c*) Vertical way (*d*) Horizontal way

Q. 9. Chicks come out earlier than 21 days in

(*a*) Rainy season (*b*) Winter season
(*c*) Summer season (*d*) Spring season

Q. 10. Oldest records of artificial incubation have been found in

(*a*) England (*b*) Egypt
(*c*) China (*d*) India

Q. 11. Optimum oxygen concentration for incubation is

(*a*) 25% (*b*) 15%
(*c*) 21% (*d*) 10%

Q. 12. Optimum CO_2 concentration for hatchability is

(*a*) 1% (*b*) 5%
(*c*) Below 0.5% (*d*) Below 0.1%

13. Optimum relative humidity for hatching is

(*a*) 61% (*b*) 70%
(*c*) 90% (*d*) None of the above

Q. 14. What is the level of CO_2 at which the hatchability becomes zero?

(*a*) 0.05% (*b*) 5%

(*c*) 0.5% (*d*) None of the above

Q. 15. The temperature in still air incubator should be at least

(*a*) 5°F

(*b*) 3°F

(*c*) 1°F above that of forced draft type incubators.

(*d*) None of the above

Q. 16. For finding out fertility percentage candling of egg is done on

(*a*) Tenth day (*b*) Seventh day

(*c*) Second day (*d*) None of the above

Q. 17. For finding out the faults of incubator the hatchability should be calculated:

(*a*) On all eggs set (*b*) On fertile eggs

(*c*) On eggs laid by the hen (*d*) None of the above

Q. 18. In Hatcher the temperature should be reduced to

(*a*) 95°F (*b*) 90°F

(*c*) 98°F (*d*) None of the above

Q. 19. Humidity is increased in the Hatcher

(*a*) As soon as the eggs are transferred to Hatcher

(*b*) When the hatching starts

(*c*) When hatching has stopped

(*d*) None of the above

Q. 20. Second candling of eggs to remove the dead embryos should be done after

(*a*) 10 days (*b*) 14 to 16 days

(*c*) 5 days of incubation (*d*) 18 days

I. C. Multiple Choice Questions (Tick Correct Answer/s)

Q. 1. **The embryonic threshold temperature of chicken is_______above which embryonic growth commences and below which ceases**

(*a*) 24°C (*b*) 20.4°C

(*c*) 19.4°C (*d*). 18.4°C

Q. 2. **The temperature in the egg holding room should be_______to curtail embryonic development completely, if the eggs are to be hold for not more than 5 days.**

(*a*) 17.3°C (*b*) 18.3°C

(*c*) 19.3°C (*d*) 20.3°C

Q. 3. **The relative humidity in the egg holding room should be_______**

(*a*) 55-60% (*b*) 65-70%

(*c*) 75-80% (*d*) 85-90%

Q. 4. **Hatching time is delayed by 30 minutes and hatchability is reduced_______% for each day eggs are held or stored beyond 4 days.**

(*a*) 1% (*b*) 2%

(*c*) 3% (*d*) 4%

Q. 5. **Better hatchability can be obtained if the Haugh Units of eggs are more than_______**

(*a*) 50 (*b*) 60

(*c*) 70 (*d*) 80

Q. 6. **The upper lethal body temperature of the bird is about**

(*a*) 47.2°C (*b*) 46.2°C

(*c*) 45.2°C (*d*) 44.2°C

Q. 7. **The ammonia concentration in the poultry house should not be more than_______ppm.**

(*a*) 50 (*b*) 25

(*c*) 30 (*d*) 45

Q. 8. At an average humidity birds will begin panting when the ambient temperature reaches_______

(*a*) 26.4°C (*b*) 27.4°C

(*c*) 28.4°C (*d*) 29.4°C

Q. 9. On an average as a rule of thumb the brown egg layer produces_______Btu of heat per hour

(*a*) 25 (*b*) 30

(*c*) 35 (*d*) 45

Q. 10. The light threshold of the birds during the growing period is about_______fc.

(*a*) 0.25 (*b*) 0.50

(*c*) 1.00 (*d*) 1.50

Q. 11. In a comfortable environment nearly_______% of the total heat produced by the birds is lost through radiation, conduction and convection.

(*a*) 50 (*b*) 60

(*c*) 70 (*d*) 75

Q. 12. In the adult chicken the body temperature variability ranges between_______to_______°F

(*a*) 99 to 101 (*b*) 101 to 103

(*c*) 103 to 105 (*d*) 105 to 107

Q. 13. When baby chicks are reared successfully without any extra heat, such brooding is known as

(*a*) Dry brooding

(*b*) Wet brooding

(*c*) Warm brooding

(*d*) Cold brooding

Q. 14. The optimum temperature for maximal rate of egg production in layers is_______°F

(*a*) 50 (*b*) 55

(*c*) 65 (*d*) 75

Q. 15. For fumigation of incubators and hatchers, potassium permanganate and formalin are used in the ratio of

(*a*) 2:1 (*b*) 3:5

(*c*) 2:5 (*d*) 1:2

Q. 16. For obtaining good hatchability eggs are usually set in the incubator

(*a*) With narrow end-up (*b*) With broad end-up

(*c*) Horizontally (*d*) All the above

Q. 17. Relatively high humidity in the Hatcher than the incubator helps in

(*a*) Reduction in post hatch mortality

(*b*) Decreases lameness

(*c*) Pipping of chickens

(*d*) All the above

Q. 18. During the process of candling of incubated eggs on 18 days of incubation they are usually classified as transparent, translucent, and opaque. The term Opaque refers to

(*a*) Unfertile eggs (*b*) Dead germs

(*c*) Eggs with live embryo (*d*) None of the above

Q. 19. Artificial insemination is usually done in the late hours of the day because

(*a*) Late hours are cooler

(*b*) The egg (Ovum) is in the right stage of formation for ensuring high fertility

(*c*) Hard shelled eggs are usually absent in the uterus

(*d*) None of the above

Q. 20. Optimum oxygen concentration for normal incubation of egg is

(*a*) 11% (*b*) 21%

(*c*) 31% (*d*) 41%

I. D. Multiple Choice Questions (Tick Correct Answer/s)

Q. 1. What is the threshold temperature for the embryonic development?

(*a*) 60°F (*b*) 65°F

(*c*) 70°F (*d*) 75°F

Q. 2. Mention the type of organisms involved in contamination of hatching eggs.

(*a*) *E. coli* and *Pseudomonas*

(*b*) *Salmonella* and *Arizona*

(*c*) Both (*a*) and (*b*)

(*d*) Neither (*a*) nor (*b*)

Q. 3. What vitamin can be produced industrially from phytic acid?

(*a*) Vitamin B_1 (*b*) Vitamin B_2

(*c*) Vitamin B_6 (*d*) Inositol

Q. 4. Incubation period in Chicken is_______days

(*a*) 18 days (*b*) 21 days

(*c*) 28 days (*d*) 35 days

Q. 5. First major peak of embryo mortality occurs at

(*a*) At day zero

(*b*) Between 2 to 4th days of setting

(*c*) Between 5 to 7th day of setting

(*d*) None of the above

Q. 6. For finding out the faults of incubation what records of hatching should be known

(*a*) No. of chicks hatched

(*b*) No. of chicks crippled

(*c*) No. of dead in shell

(*d*) All three

Q. 7. The fertility of the eggs is known through

(*a*) The chicks hatched

(*b*) The chicks died in shell found in second candling

(*c*) The eggs clear found at first candling

(*d*) None of the above

Q. 8. What is a physiological zero?

(*a*) Temperature at which the birds survive

(*b*) Temperature at which embryonic development starts

(*c*) Temperature at which the birds feels comfortable

(*d*) None of the above

Q. 9. In the event of power failures cooling takes place if the doors of the incubator are open does it affects

(*a*) Hatchability

(*b*) Embryonic mortality

(*c*) Increase in the number of deformed or week chicks

(*d*) None of the above

Q. 10. When the temperature is 106 °F for at least 3 hours it causes

(*a*) Increase embryonic mortality

(*b*) Does not affect hatchability

(*c*) Does affect fertility of eggs

(*d*) None of the above

II. Fill in the Blanks in the Following Sentences

1. The follicle stimulating hormone (FSH) is produced by_______

2. Laid down on the outside of the shell in the uterus is the last of the concentric layers of egg formation is known as_______

3. Pigments produced in the uterus at the time when the shell is produced are responsible for_______

4. The______is the albumen-secreting portion of the oviduct.

5. During the course of a normal mating between male and female chicken, the male ejaculate from ______to______billion sperm cells.

6. Under good management practice; one broiler breeder hen gives______number of chicks.

7. Best method of disinfection of hatching eggs is through______

8. Eggs should be set in the standing position keeping the broader end at the______

9. Hatching eggs should be______during storage also.

10. During storage hatching eggs should be stored at a temperature below______°F.

11. Cracked hatching eggs should be______before setting for hatching.

12. The loss of viscosity of the albumen is accompanied by the decrease in______, a protein essential for the embryonic growth.

13. Temperature in the egg holding room should be______to curtail embryonic development completely, if the eggs are to be hold for not more than 5 days.

14. Better hatchability can be obtained if the Haugh units of eggs are more than______

15. For fumigation two parts (by volume) of______are mixed with approximately one part (by wt.) of______.

16. For every percent reduction in oxygen content below 21%, there is decrease of______% hatchability and for every percent increase in oxygen content above to this hatchability decreases by______percent.

III. Mark the Following Statements as True or False

1. When the egg is first laid it contains small air cell.
2. Hatching eggs from a young breeder flock can be held longer than those from an old breeder flock.
3. Moisture laden eggs can be only fumigated with formaldehyde gas.
4. Reducing the humidity of the incubating compartments shortens the hatching period and increasing it lengths the incubation time.
5. Eggs of large size (above 60 g) show better hatchability.
6. Thin shelled eggs hatch better than the thick shelled eggs.
7. Storing the hatching eggs up to 10 days does not have any adverse effect on hatchability.
8. Physiological zero temperature in case of hen egg ranges between 20 and 21 °C.
9. Quails have longer incubation period than chicken.
10. For normal hatching of eggs carbon dioxide concentration in the incubator should not be more than 2 percent.
11. For better hatchability eggs are set with large and up.
12. Incubated eggs should be tested for fertility only once on 5th to 7th days.
13. Fertility is moderately heritable trait.
14. The most popular disinfectants used in the hatchery operation are the phenolic compounds.
15. The vent method of sexing is the most popular and it was originally developed in Japan.
16. None of the chicks should be taken out of the incubation machine before they are fluffed out.

17. Lower the reading of Haugh units, the better the hatchability of the eggs.

18. Cooling incubating eggs during the first 19 days of incubation decreases embryonic malpositions.

19. Decreasing the humidity of the setter lengthens the incubation period.

20. Smaller eggs lose a larger percentage of their weight during incubation than larger eggs.

21. In the Hatcher, the female chicks hatch earlier than their corresponding male chicks.

22. Leghorn eggs have a shorter incubation period than those produced by meat type birds.

23. Incubating the eggs large end up helps in reducing the malpositions.

24. Shell quality of the eggs affects the humidity requirement in the setter.

Answers

I. A. Multiple Choice Questions

1	(*a*)	6	(*c*)	11	(*a*)	16	(*a*)
2	(*c*)	7	(*b*)	12	(*b*)	17	(*a*)
3	(*b*)	8	(*a*)	13	(*c*)	18	(*c*)
4	(*c*)	9	(*c*)	14	(*a*)	19	(*a*)
5	(*a*)	10	(*a*)	15	(*b*)		

I. B. Multiple Choice Questions

1	(*a*)	6	(*d*)	11	(*c*)	16	(*b*)
2	(*a*)	7	(*a*)	12	(*c*)	17	(*b*)
3	(*b*)	8	(*a*)	13	(*a*)	18	(*c*)
4	(*b*)	9	(*c*)	14	(*b*)	19	(*a*)
5	(*b*)	10	(*c*)	15	(*c*)	20	(*d*)

I. C. Multiple Choice Questions

1	(*a*)	6	(*a*)	11	(*d*)	16	(*b*)
2	(*b*)	7	(*b*)	12	(*d*)	17	(*d*)
3	(*c*)	8	(*d*)	13	(*d*)	18	(*c*)
4	(*d*)	9	(*d*)	14	(*d*)	19	(*c*)
5	(*d*)	10	(*b*)	15	(*d*)	20	(*b*)

I. D. Multiple Choice Questions

1	(*d*)	4	(*b*)	7	(*c*)	10	(*a*)
2	(*c*)	5	(*b*)	8	(*b*)		
3	(*d*)	6	(*d*)	9	(*c*)		

II. Fill in the Blanks

1. Anterior pituitary gland.
2. Cuticle
3. Colour
4. Magnum
5. 1.5-8 billion
6. About 150 Nos.
7. Fumigation
8. Top
9. Turned
10. 65
11. Removed
12. Ovomucin
13. 65 °F or 18.3 °C
14. 80
15. Formalin, Potassium permanganate
16. 5% 1%

III. Mark the Following Statements as True or False

1	False	8	True	15	True	22	True
2	True	9	False	16	True	23	True
3	False	10	False	17	False	24	True
4	False	11	True	18	False		
5	False	12	True	19	False		
6	False	13	True	20	True		
7	False	14	False	21	True		

Chapter 12

Poultry Housing and Management

I. A. Multiple Choice Questions (Tick Correct Option/s)

Q. 1. During the first two weeks of brooding period the temperature of the brooder house should be

(a) 85 °F (b) 95 °F

(c) 75 °F (d) 65

Q. 2. The ammonia level produced by litter should not exceed

(a) 55 ppm (b) 45 ppm

(c) 25 ppm (d) 35 ppm

Q. 3. The total air breathed by fowl at rest in comfortable environment_______

(a) 1.6 cubic foot per hour/ bird

(b) 1.8 cubic foot per hour/ bird

(c) 2.0 cubic foot per hour/ bird

(d) 2.2 cubic foot per hour/ bird

Q. 4. The layers are provided for artificial light

(*a*) 12 hours a day (*b*) 14 hours a day

(*c*) 16 hours a day (*d*) 18 hours a day

Q. 5. The ratio of males to female for Rhode Island Red breed should be

(*a*) 1 to 8 (*b*) 1 to 10

(*c*) 1 to 15 (*d*) 1 to 20

Q. 6. Out door laying cages are popular where

(*a*) Little protection from cold is needed

(*b*) Protection from cold is not needed

(*c*) Little protection from sun is needed

(*d*) None of the above conditions

Q. 7. The broiler chicks have the inherited ability for

(*a*) Better feed efficiency (*b*) Rapid growth

(*c*) Less mortality (*d*) Cannibalism

Q. 8. The most important point in selecting the material of deep litter is

(*a*) It should be easily available

(*b*) The cost should be low

(*c*) It should absorb the moisture

(*d*) None of the above

Q. 9. Breeders and hatcheries have to do isolation and sanitation program to maintain

(*a*) Mycoplasma free flock (*b*) A.L.C. free flock

(*c*) Disease free flock (*d*) High producing flock

Q. 10. A flock of 100 birds would produce fresh manure in one year as below

(*a*) 3 tons (*b*) 7 tons

(*c*) 10 tons (*d*) 13 tons

Q. 11. The order in which the pigment disappears in laying hen

(*a*) Eye Ring, Vent, Beak

(*b*) Beak, Vent, Eye ring

(*c*) Vent, Beak, Eye ring

(*d*) None of the above

Q. 12. When baby chicks are reared successfully without any extra heat, such brooding is known as

(*a*) Cold brooding (*b*) Hot brooding

(*c*) Dry brooding (*d*) Wet brooding

Q. 13. The floor space requirement up to tenth week is

(*a*) 45 sq inches per chick

(*b*) 55 sq inches per chick

(*c*) 75 sq inches per chick

(*d*) 65 sq inches per chick

Q. 14. The optimal temperature for maximum rate of egg production is

(*a*) 45 °F (*b*) 65 °F

(*c*) 75 °F (*d*) 85 °F

Q. 15. Old Built up litter contains fewer yeasts, molds and bacteria because of its increased

(*a*) Alkalinity (*b*) Acidity

(*c*) Salinity (*d*) None of the above

Q. 16. The broiler finishing diet should contain increased level of

(*a*) Protein and Xanthyhill

(*b*) Energy and Xanthyhill

(*c*) Calcium and Xanthyhill

(*d*) Fiber and Xanthyhill

Q. 17. The depth of litter material should be

(*a*) 9 to 10 inches (*b*) 6 to 8 inches

(*c*) 5 to 7 inches (*d*) 2 to 4 inches

Q. 18. During the first 12 hours of incubation eggs are fumigated with

(*a*) 1 X concentration of $KMnO_4$ and formaldehyde

(*b*) 2 X concentration of $KMnO_4$ and formaldehyde

(*c*) 3 X concentration of $KMnO_4$ and formaldehyde

(*d*) 4 X concentration of $KMnO_4$ and formaldehyde

Q. 19. As compared to other domestic animals the poultry manure is much richer in

(*a*) Nitrogen (*b*) Potash

(*c*) Phosphorus (*d*) All of the above

Q. 20. Important factors in distinguishing between good and poor layers during first annual molting are

(*a*) Temperature and duration (*b*) Time and duration

(*c*) Humidity and duration (*d*) None of the above

I. B. Multiple Choice Questions (Tick Correct Option/s)

Q. 1. A comfortable poultry house must provide

(*a*) Adequate space (*b*) Supply fresh air

(*c*) Remain dry (*d*) All the above

Q. 2. A poultry house may be

(*a*) Economical (*b*) Durable

(*c*) Provide comfort and safely (*d*) All the above

Q. 3. The location of poultry house should be

(*a*) In the vicinity of town (*b*) Easily accessible

(*c*) Good soil (*d*) None of the above

Q. 4. The floor space for layer birds depend upon

(*a*) Type of floor (*b*) Size of birds

(*c*) Temperature and ventilation

(*d*) All the above

Q. 5. Building large poultry pens reduce

(*a*) Housing cost (*b*) Labor cost

(*c*) Gathering of eggs (*d*) All the above

Q. 6. Egg production and feed consumption is influenced by housing temperature at 75°F.

(*a*) Maximum (*b*) Minimum

(*c*) All the above (*d*) None of the above

Q. 7. Brooder house should be

(*a*) Isolated (*b*) Irruption

(*c*) Irrigated (*d*) Introverted

Q. 8. The floor space requirement per bird (square meter) during the brooding period in leghorn may be

(*a*) 0.07 (*b*) 0.093

(*c*) 0.116 (*d*) 0.136

Q. 9. The type of floor in the growing house may be

(*a*) Littered floor (*b*) Slat and litter floor

(*c*) All-slat floor (*d*) Any one of above

Q. 10. Many cage arrangements are recommended to get more cages in a given area which is

(*a*) Single deck (*b*) Double deck

(*c*) Triple deck (*d*) All the above

Q. 11. The commercial chicken production manual which is written by Mack-O-North, is published by

(*a*) Vikas Publishing company

(*b*) The AVI Publishing company

(*c*) Tata McGraw Hill Publishing Company

(*d*) Oxford and IBH Publishing Company

Q. 12. The bird has self regulatory process by which it tends to keep its own temperature within certain limits which is known as

(*a*) Homeostasis (*b*) Homeopathy

(*c*) Homologue (*d*) Homonymy

Q. 13. The method of heat liberation from the body of birds is by

(*a*) Radiation (*b*) Conduction

(*c*) Convention (*d*) All the above

Q. 14. In hot weather the cooling of open sided bird houses are done by

(*a*) Sprinkle of house roof

(*b*) Sprinkle the ground area

(*c*) Fans

(*d*) All the above

Q. 15. A controlled environment house is one in which inside conditions are maintained as near as possible to the birds optimum conditions which are done by

(*a*) Cooling and heating system (*b*) Artificial light

(*c*) All the above (*d*) None of the above

Q. 16. In hot tropical climate, the long axis of poultry houses is usually oriented

(*a*) From North to South (*b*) From East to West

(*c*) Both (*a*) and (*b*) (*d*) None of the above

Q. 17. Rearing birds in cages has many advantages such as

(*a*) Economy in space

(*b*) Prevention and control of certain diseases like Coccidiosis

(*c*) Allows better flock supervision

(*d*) All the above

Q. 18 Light stimulates pituitary gland through optic nerve hence, as a 'Rule of Thumb' the growing pullets should be subjected to

(*a*) Increasing amount of light

(*b*) Decreasing amount of light

(*c*) Equal amount of light and darkness

(*d*) None of the above

Q. 19. The advantage of growing male and female broilers separately is

(*a*) Ensures even growth and reduces culls

(*b*) Improves feed efficiency

(*c*) Both (*a*) and (*b*)

(*d*) None of the above

Q. 20. Restricted feeding of broiler parent stock during growing stage helps

(*a*) To reduce the body size of adults males & females

(*b*) Reduce feed consumption and improved economy

(*c*) Improves production and reproduction of the parents stock

(*d*) All the above

Q. 21. Increasing day length results in

(*a*) Release of gonadotrophic hormones

(*b*) Check the release of gonadotrophic hormones

(*c*) Have no effect on release of gonadotrophic hormones

(*d*) None of the above

Q. 22. To use auto sex determination in sex, what combination of parents should be used?

(*a*) Non barred male and barred female

(*b*) Barred male × non barred female

(*c*) Barred male × barred female

(*d*) None of the above

Q. 23. The molting order in laying hen is

(*a*) Head, Neck, Breast, Body, Wings, and Tail

(*b*) Tail, Wings, Body, Breast, Neck and Head

(*c*) Body, Breast, Wings, Neck, Tail and Head

(*d*) Breast, Body, Wings, Neck, Head and Tail

Q. 24. The ratio of body length to intestinal length in chicken is

(*a*) 1:4 (*b*) 1:2

(*c*) 1:3 (*d*) 1:6

Q. 25. When the temperature is optimum for the chickens in the brooder house

(*a*) The chicken are closer to the source of heat

(*b*) Huddle down under brooder

(*c*) Move away from the source of heat

(*d*) Evenly scattered in the house

Q. 26. Built-up litter is a rich source of

(*a*) Vitamin B_1 (*b*) Vitamin B_{12}

(*c*) Vitamin D (*d*) Vitamin E

II. Fill in the Blanks

1. Unless means are provided to remove it, moisture will_______in the poultry house.
2. For heat removal from poultry houses 50 cu. ft. of air will remove 1 Btu of heat for each_______rise in temperature.
3. In order to get maximum laying response from the layer light programme, the total daily light should be_______hours light day.
4. The optimum layer house temperature should be between_______.
5. It is estimated that from_______% of all the commercial layers in the world are kept in cages.
6. The upper lethal body temperature of the bird is about_______
7. The ammonia concentration in the poultry house should not be more than_______ppm.

8. When the temperature at the bird's surface is greater than the adjacent air, the heat is lost from the body by______

9. At an average humidity birds will begin panting when the ambient temperature reaches_______

10. On an average as a rule-of-thumb the brown egg layer produces_______Btu of heat per hour.

11. In layers for optimum egg production the light should be provided for more than_______hrs per day.

12. The light threshold of the birds during the growing period is about_______%.

13. In a comfortable environment nearly_______% of the total heat produced by the birds is lost through radiation, conduction and convection.

14. The width of the open sided poultry house should not be more than_______ft.

15. In the adult chicken the body temperature variability ranges between_______° to_______°.

16. When baby chicks are reared successfully without any extra heat, such brooding is known as_______.

17. The floor requirement up to tenth week is_______sq inches per chick.

18. The optimal temperature for maximal rate of egg production in layer is_______.

19. Breeding layers need relatively more_______,_______and_______in their ration for better hatchability.

20. Old built up litter contains fewer yeasts, moulds and bacteria because of its increased_______.

21. The broiler finishing diet should contain increased level of_______and_______pigment.

22. The depth of the litter material should be_______inches at initials stage.

23. Eggs are fumigated with_______during the first 12 hrs of incubation.

24. Poultry manure is much richer in_______than that of other domestic animals.

25. Time and duration of first annual moult are important point in distinguishing between_______and _______ layer.

26. The cage floor space allowance in commercial practices for WL chicken is_______

III. Write True or False Against Following Statements

1. An all-in, all-out brooding programme is recommended.
2. For brooder houses sparrow-proofing is not essential.
3. Fresh air is needed in the bird houses but no draft.
4. Continuous ridge ventilators are better in layer houses.
5. Usually about 50% of all heat generated by the bird is lost through radiation, conduction and convention.
6. The rate of loss of body heat is influenced by ambient temperature.
7. Panting means bringing more outside air in contact with the membrane of respiratory tract.
8. For each 1% increase in the temperature of the layer house the heat dissipation decreases by 0.5%.
9. There are many materials used to insulate poultry houses which are considered on their thermal resistance.

10. Moisture in the poultry houses constitutes less problem of adequate poultry housing.

11. The height of the house or stud length should be 8 to 10.

12. Practically all poultry houses built today have no gable roof.

13. For disease control programme, a concrete floor is mandatory in bird houses.

14. Houses equipped with cages necessitate the greatest amount of air movement than other type of houses.

15. Pullets may be transferred to the laying cages at any age between 14 to 22 weeks.

16. The temperature of egg cooling room should be more than 20°C.

17. In many locations damp-proof concrete floors are a pre-requisite of good house construction.

18. Unless the poultry building is extremely wide, the truss design should be used for supporting the roof.

19. Inside of the walls of poultry houses should not be covered with a glazed, hard, non-absorbent finish.

20. Correct poultry housing is a prerequisite of any good programme of poultry management.

21. Smaller breeds have a lower body temperature than larger breeds.

22. Chickens being homeothermic have ability to maintain a rather uniform temperature of their internal organs within certain limits.

23. At an average humidity birds will begin panting when the ambient temperature reached 85°F.

24. Insensible heat loss from the body of the birds increases the temperature of the surroundings.

25. Practically all poultry houses built today have preferably the gable roof.

26. Beak trimming helps to prevent cannibalism in a flock.

27. Feathering appears to be better in the birds kept in cages as compared to those maintained on litter floor.

28. Egg shell quality deteriorates rapidly when layers are kept in cages.

29. Usually early moulters are good layers.

30. All chickens have colour vision.

31. The temperature of a newly hatched chick is about 107 °F.

32. At 70 °F the chicken normally drink twice the weight of water compared to the feed consumed.

33. When the relative humidity of the air in the house is 50% the moisture content of the litter is usually approaching 25%

34. Birds kept on wire consume more water than those kept on the floor.

35. Usually a temperature of 85° to 90°F outside the canopy is satisfactory for day old chicks reared on litter floor.

36. Houses equipped with cages necessitate the lesser amount of air movement as compared to the litter houses.

37. As a general rule roosts are not needed in a poultry house either for growing or laying birds.

38. The use of artificial light in the laying house is mandatory.

39. Incidences of broken eggs are much more in the layers kept on litter floor.

40. Probably the biggest reason for keeping a required percentage of moisture in the litter has to do with the control of coccidiosis.

41. The temperature of the brooder house should be 95°F during the first two weeks of brooding period.

42. The daily water requirement of 100 laying bird is 3 gallons.

43. Abrupt change in feed in laying hens results in fall in consumption of feed and corresponding loss in production.

44. Pullets should be fed a diet containing 15 to 16 per cent protein

45. Trap nests are essential equipment for pedigree breeding, where one female is continuously mated to one female in a breeding pen.

46. Individual laying cages are mainly used to solve the cannibalism in layers.

47. The purpose of caponizing is to improve the quality of Poultry Meat.

48. The desirable pH for ideal litter in Poultry is 8.0 or more.

49. Fore preserving the fertilizer value of poultry manure Di-Ammonium Phosphate should be added.

50. The moulting order in laying hen is head, neck, breast, body, wings and tail.

51. For best results, moulting should be done after 40 weeks of lay.

52. In hot weather caged layer suffer more severely than birds on deep litter.

53. Sex-differentiation at day-old by Japanese method is called auto-sexing.

54. Bigger poultry house have more usable space per bird as compared to smaller house.

IV. Match the Following

1.	Battery brooder	(*a*)	Saw dust
2.	Mating system	(*b*)	Cracked egg.
3.	Deep litter	(*c*)	Poultry Manure
4.	Demerit of cage system	(*d*)	Curled toe paralysis
5.	Super phosphate	(*e*)	Conclusion of one year production
6.	Riboflavin	(*f*)	Nervous excitement
7.	Trap nest	(*g*)	Artificial Brooding
8.	Male Female Ratio	(*h*)	Artificial Insemination
9.	Moulting	(*i*)	Record keeping
10.	Lightening	(*j*)	Hatchability

Answers

I. A. Multiple Choice Questions

1	(*b*)	6	(*a*)	11	(*d*)	16	(*b*)
2	(*c*)	7	(*b*)	12	(*a*)	17	(*d*)
3	(*a*)	8	(*c*)	13	(*b*)	18	(*b*)
4	(*c*)	9	(*a*)	14	(*c*)	19	(*a*)
5	(*b*)	10	(*b*)	15	(*a*)	20	(*b*)

I. B. Multiple Choice Questions

1	(*d*)	8	(*a*)	15	(*c*)	22	(*a*)
2	(*d*)	9	(*d*)	16	(*b*)	23	(*a*)
3	(*d*)	10	(*c*)	17	(*d*)	24	(*a*)
4	(*d*)	11	(*b*)	18	(*b*)	25	(*d*)
5	(*d*)	12	(*a*)	19	(*c*)	26	(*b*)
6	(*a*)	13	(*d*)	20	(*d*)		
7	(*a*)	14	(*d*)	21	(*a*)		

II. Answer Key for Fill in the Blanks

1. Increase
2. 1 °F to 0.55 °F
3. 14-16 hours light day
4. 65-76 °F or 18.3- 23.9 °C
5. 60-65%

6. 116.8 °F or 47°C
7. 25
8. Radiation
9. 86 °F or 29.4 °C
10. 45
11. 14
12. 0.5
13. 75
14. 40
15. 105 °F to 107 °F or 40.6 to 41.7 °C
16. Cold brooding
17. 55
18. 75 °F
19. riboflavin, pantothenic acid, biotin and manganese
20. alkalinity
21. fat, xanthophylls
22. 2 to 4
23. formaldehyde gas
24. nitrogen
25. poor, good
26. About 60 sq. inches

III. Answer Key for True or False Statements

1.	True	8.	True	15.	True	22.	True
2.	False	9.	True	16.	False	23.	True
3.	True	10.	False	17.	True	24.	False
4.	True	11.	True	18.	True	25.	True
5.	False	12.	False	19.	False	26.	True
6.	True	13.	True	20.	True	27.	False
7.	True	14.	True	21.	False	28.	True

29.	**False**	**36.**	**False**	**43.**	**True**	**50.**	**True**
30.	**True**	**37.**	**True**	**44.**	**True**	**51.**	**False**
31.	**False**	**38.**	**True**	**45.**	**True**	**52.**	**True**
32.	**True**	**39.**	**False**	**46.**	**False**	**53.**	**False**
33.	**True**	**40.**	**True**	**47.**	**True**	**54.**	**True**
34.	**True**	**41.**	**True**	**48.**	**False**		
35.	**True**	**42.**	**False**	**49.**	**False**		

III. Answer Key Match the Following

1.	(*g*)	4.	(*b*)	7.	(*i*)	10.	(*f*)
2.	(*h*)	5.	(*c*)	8.	(*j*)		
3.	(*a*)	6.	(*d*)	9.	(*e*)		

Chapter 13

Poultry Feeding and Nutrition

I. A. Multiple Choice Questions (Tick the Correct Answer/s)

Q. 1. When chicks are fed raw-egg white, they develop deficiency of

(*a*) Choline (*b*) Pyridoxine

(*c*) Biotin (*d*) Pantothenic acid

Q. 2. The intensive rearing of poultry became popular only after the discovery of

(*a*) Vitamin A (*b*) Vitamin D

(*c*) Vitamin C (*d*) Vitamin B_2

Q. 3. The classes of protein found in yolk according to their behavior in ultracentrifuge are

(*a*) Low density (*b*) Water soluble

(*c*) Granular fractions (*d*) All of these

Q. 4. Hemorrhagic syndrome observed in growing chick might be due to

(*a*) Deficiency of Vitamin K

(*b*) Injudicious use of sulfaquinoxaline

(*c*) Molds

(*d*) All the above

Q. 5. The intensive type of poultry production useful to those who have small quantity of floor space at their disposal is

(*a*) Deep litter system (*b*) Battery system

(*c*) Folding unit system (*d*) All the above

Q. 6. The most abundant protein in egg white is

(*a*) Ovoglobulin (*b*) Ovalbumin

(*c*) Ovomucoid (*d*) Ovomucin

Q. 7. The most forgotten nutrient in poultry is

(*a*) Fishmeal (*b*) Water

(*c*) Vitamin (*d*) Maize

Q. 8. Forced feeding is some times attempted in the first week in

(*a*) Chicks (*b*) Turkey

(*c*) Duck (*d*) Quails

Q. 9. Availability of phosphorous is poor from

(*a*) Animal Source (*b*) Inorganic Source

(*c*) Plant Source (*d*) All the above

Q. 10. The normal position of embryo during hatching is

(*a*) Head under right wing in narrow end of egg

(*b*) Head under left wing in broad end of egg

(*c*) Head under right wing in broad end of egg

(*d*) None of the above

Q. 11. Abnormal embryo position may be caused by

(*a*) Inheritance (*b*) Balanced ration

(*c*) Normal incubation (*d*) All the above

Q. 12. If temperature in dry and wet bulb reading is same, it indicates

(*a*) Low humidity

(*b*) No water in the dry bulb

(*c*) No water in the bottle of wet bulb

(*d*) None of the above

Q. 13. Feather protein contains high proportion of

(*a*) Methionine (*b*) Cystine

(*c*) Lysine (*d*) Histidine

Q. 14. Feed conversion efficiency of layers is influenced by

(*a*) Metabolizable energy content of the diet

(*b*) Pen temperature

(*c*) Rate of egg production

(*d*) All the above

I. B. Multiple Choice Questions (Tick the Correct Answer/s)

Q. 1. More defined poultry nutrients requirement are known in respect of

(*a*) Layers only

(*b*) Broilers only

(*c*) Broiler and layers both

(*d*) None of these

Q. 2. More defined nutrients requirements in poultry could be possible due to

(*a*) Nature of the bird

(*b*) (*a*) + specific conditions of poultry rearing

(*c*) (*b*) + short period of involvement

(*d*) None of these

Q. 3. Best animal protein source in poultry ration is

(*a*) Meat meal (*b*) Jawla fish

(*c*) Feather meal (*d*) None of these

Q. 4. Molasses and neutral fats are added to the poultry ration, However, their inclusion should not exceed

(*a*) 1-2% (*b*) 3-4%

(*c*) 5-10% (*d*) 20-30%

Q. 5. Egg shell is deposited in

(*a*) Infundibulum (*b*) Magnum

(*c*) Uterus (*d*) Isthmus

Q. 6. Another name for crazy chick disease is

(*a*) Encephalomalacia (*b*) Liver necrosis

(*c*) Diatheses (*d*) Muscular dystrophy

Q. 7. Alfalfa meal in poultry ration is used for supplying of

(*a*) Vitamins only

(*b*) Vitamins and protein

(*c*) (*b*) + Yolk pigment

(*d*) None of these

Q. 8. Poultry achieve their optimum egg production at the age of

(*a*) 18–20 weeks (*b*) 20–24 weeks

(*c*) 28–36 weeks (*d*) 56–60 weeks

Q. 9. Rice bran, rice polish and deoiled rice polish can be incorporated in poultry diets up to the extent of

(*a*) 10–20, 25–40, 10–20% respectively

(*b*) 25–40, 40–70, 1–2% respectively

(*c*) 1–3, 0.5–2, 30–40%, respectively

(*d*) None of these

Q. 10. Poultry utilizes Vitamin D more efficiently in the form of

(*a*) Cholecalciferol/D_3 (*b*) Ergocalciferol/D_2

(*c*) Both (*a*) + (*b*) (*d*) None of the avove

Q. 11. Which variety of sorghum is most suitable to include in the poultry ration?

(*a*) Dark (*b*) Green

(*c*) White (*d*) Red

Q. 12. Curled to paralysis is caused due to deficiency of Vitamin

(*a*) Vitamin B_{12}

(*b*) Vitamin B_2

(*c*) Vitamin B_6

(*d*) Pantothenic acid

Q. 13. Xeropthalmia is caused due to the deficiency of vitamin

(*a*) Vitamin A (*b*) Vitamin D

(*c*) Vitamin K (*d*) Vitamin E

Q. 14. Alfalfa meal should not exceed more than_______in poultry ration

(*a*) 10% (*b*) 20%

(*c*) 30% (*d*) 5%

Q. 15. Linseed meal prove toxic in poultry if included in the poultry ration more than

(*a*) 2–3% (*b*) 4–5%

(*c*) 6–7% (*d*) 8–10%

Q. 16. Cotton seed meal could not replace more than 50% of the Soyabean meal in grower poultry due to the presence of toxic factors like

(*a*) Gossypols and Tannins

(*b*) Gossypols and Sterculic acid

(*c*) Gossypols and Saponins

(*d*) Gossypols and Linatin

Q. 17. Canola meal is used as source of_______in poultry ration

(*a*) Energy (*b*) Vitamins

(*c*) Minerals (*d*) Protein

Q. 18. **Canola meal is not recommended in layers which lay_______shelled eggs because they are unable to oxidize TMA**

(*a*) White (*b*) Tinted

(*c*) Brown (*d*) None of these

Q. 19. **In poultry production about_______% of cost incurs on feed alone**

(*a*) 50% (*b*) 60%

(*c*) 65% (*d*) 75%

20. **How much percentage of total feed requirement is needed only for the maintenance of the poultry**

(*a*) 50% (*b*) 60%

(*c*) 65% (*d*) 75%

I. C. Multiple Choice Questions (Tick the Correct Answer/s)

Q. 1. **Protein requirement of chickens from 0 to 8 weeks of age in percentage**

(*a*) 8 (*b*) 20

(*c*) 38 (*d*) 68

Q. 2. **Vitamin A requirements of laying chickens in International units per kg of mash**

(*a*) 1000 (*b*) 4000

(*c*) 8000 (*d*) 10,000

Q. 3. **Vitamin D requirements of breeding hen in International chick units per kg of mash**

(*a*) 100 (*b*) 200

(*c*) 400 (*d*) 1200

Q. 4. **Calcium requirements of breeding hen in ration in percentage**

(*a*) 0.5 (*b*) 3.0

(*c*) 10.0 (*d*) 20.0

Q. 5. Phosphorous requirements for laying chicks in percentage

(*a*) 0.5 (*b*) 0.62

(*c*) 1.0 (*d*) 4.0

Q. 6. Calcium requirements of laying chicks in percentage

(*a*) 3.0 (*b*) 6.4

(*c*) 16.4 (*d*) 20.4

Q. 7. Protein requirements of laying chicks in percentage

(*a*) 4.5 (*b*) 8.5

(*c*) 10.5 (*d*) 18.00

Q. 8. Protein requirements for chickens of 14 to 20 weeks of age in percentage

(*a*) 16 (*b*) 24

(*c*) 36 (*d*) 48

Q. 9. Calcium requirements of chickens from Birth to 6 weeks of age in percentage

(*a*) 0.1 (*b*) 0.3

(*c*) 0.8 (*d*) 1.0

Q. 10. Phosphorous requirements of chickens from Birth to 6 weeks of age in percentage

(*a*) 0.1 (*b*) 0.2

(*c*) 0.3 (*d*) 0.5

Q. 11. 'Curled toe paralysis' occurs as a result of deficiency of

(*a*) Thiamine (*b*) Riboflavin

(*c*) Panthothenic acid (*d*) Niacin

Q. 12. 'Slipped tendon' occurs as a result of deficiency of

(*a*) Manganese (*b*) Zinc

(*c*) Vitamin A (*d*) Iron

Q. 13. 'Anaemia' occurs as a result of deficiency of

(*a*) Sodium (*b*) Potassium

(*c*) Chlorine (*d*) Iron

Q. 14. 'Exudative diathesis' occurs as a result of deficiency of

(*a*) Calcium (*b*) Phosphorous

(*c*) Selenium (*d*) Fluorine

Q. 15. 'Blood clotting' is difficult in absence of

(*a*) Calcium (*b*) Sodium

(*c*) Potassium (*d*) Magnesium

Q. 16. 'Phytates' retards the utilization of

(*a*) Sodium (*b*) Selenium

(*c*) Calcium (*d*) Boron

Q. 17. In a compounded balanced ration, the amount of bone meal is

(*a*) 0.5 kg (*b*) 10 kg

(*c*) 20 kg (*d*) 15 kg

Q. 18. Size of the egg is not affected because of

(*a*) Molybdenum (*b*) Protein

(*c*) Linoleic acid (*d*) Calcium

Q. 19. Ground nut meal contains ample amount of

(*a*) Arginine (*b*) Lysine

(*c*) Methionine (*d*) None of the above

Q. 20. Protein requirements for breeding duck in percentage is

(*a*) 10 (*b*) 12

(*c*) 15 (*d*) 20

I. D. Multiple Choice Questions (Tick the Correct Answer/s)

Q. 1. Laying period in white leg horn may be divided into phase I up to 42 weeks and phase II from 42 weeks to the end of production. The dietary protein requirement (%) remains:

(*a*) Constant in both the phases

(*b*) Higher in first phase

(*c*) Higher in second phase

(*d*) Low in first phase

Q. 2. Within phase stated above the dietary protein requirement (%) remains

(*a*) Low in hot climate

(*b*) High in hot climate than the moderate climate

(*c*) Remains constant in both types of climates

(*d*) High in moderate climate

Q. 3. If Protein requirement is 17 g/d/hen and estimated energy requirement is 305 Kcal ME /d/hen in moderate climate then the requirement of energy in hot climate in first phase stated in (1) will be:

(*a*) Lower than 305 Kcal ME

(*b*) Higher than 305 Kcal ME

(*c*) Exact 305 Kcal ME

(*d*) Double of 305 Kcal ME

Q. 4. Nutrients requirements of broiler and layers

(*a*) Not exactly defined

(*b*) Are quite variable

(*c*) More precisely defined

(*d*) Are known only for energy

Q. 5. During feed formulation special provision of allowances are made in respect of certain nutrients because of

(*a*) Their destruction during feed processing and storage

(*b*) Lesser utilization of the nutrient at cellular level

(*c*) Due to both above reason

(*d*) None of the above

Q. 6. Birds eat to satisfy their energy need because energy is

(*a*) Not a prime nutrient

(*b*) Last nutrient utilized in low amounts

(*c*) A prime nutrient utilized in largest amounts

(*d*) Not essential for the vital function

Q. 7. Nutrient requirements in broiler and layer are well defined because of

(*a*) The nature of birds

(*b*) The short periods of time involved

(*c*) The specific conditions under which they are produced

(*d*) The above all reasons

Q. 8. Starter ration of chicken contains 20-23% protein whereas protein% in starter ration of turkeys, quail, pheasants and partridge is

(*a*) 13% (*b*) 16%

(*c*) 19% (*d*) 28%

Q. 9. Suggested protein requirement in Turkey layer breeder ration is

(*a*) 14% (*b*) 18%

(*c*) 22% (*d*) 26%

Q. 10. Allowance factors are used for vitamins in ration formulations, the suggested allowance factor for Vitamin A is

(*a*) 10% (*b*) 25%

(*c*) 40% (*d*) 55%

Q. 11. The suggested allowance factor for water soluble (B complex) Vitamins is about

(*a*) One third of Vitamin A (*b*) Equal to Vitamin A

(*c*) One half of Vitamin A (*d*) Zero

Q. 12. Suggested fat contents in broiler breeding flock ration is

(*a*) 1% (*b*) 2%

(*c*) 3% (*d*) 4%

Q. 13. Broiler starter ration should contain 1% lysine which in breeders should be reduced to

(*a*) 0.8% (*b*) 0.5%

(*c*) 0.3% (*d*) 0.1%

Q. 14. In comparison to starter ration pre-breeders and breeders flock should contain

(*a*) Low Vitamin A, D & E

(*b*) Same content of Vitamin A, D & L

(*c*) High Vitamin A, D & E

(*d*) None of these Vitamins

Q. 15. Energy requirements in birds diet depends upon the

(*a*) Nature of the feed

(*b*) Regulation of feed intake

(*c*) True digestibility and metabolism of the feed nutrient in the birds

(*d*) All above factors

Q. 16. Ducklings between 0 to 4 week of age do not show variable response to different dietary energy level within limits because they

(*a*) Adjust more efficiently to their ME intake

(*b*) Do not eat more of energy

(*c*) Cannot adjust to ME intake

(*d*) Eat more energy

Q. 17. In Japanese laying quail diet optimum level of Mn seems to be

(*a*) 20 mg/kg (*b*) 40 mg/kg

(*c*) 60 mg/kg (*d*) 80 mg/kg feed

Q. 18. In Japanese quail for egg laying the optimum dietary concentration of calcium, seems to be

(*a*) 1% (*b*) 2%

(*c*) 3% (*d*) 4%

Q. 19. The suggested Zinc concentration in the ration of broiler breeding flock is

(*a*) 20 ppm (*b*) 40 ppm

(*c*) 60 ppm (*d*) 80 ppm

Q. 20. Energy intake of the birds are governed by the environmental temperature but growth and egg production need specific amount of the nutrients and therefore the concentration of nutrients in the diet other than the energy is

(*a*) Reduced along with energy at higher temperature

(*b*) Remains constant with energy

(*c*) Increased in relation to energy at higher temperature

(*d*) Not known exactly

I. E. Multiple Choice Questions (Tick the Correct Answer/s)

Q. 1. At environmental temperature above 30°C voluntary feed intake

(*a*) Increases (*b*) Decreases

(*c*) No change

Q. 2. If a higher concentration of energy is used in the diet of chick, feed consumption will

(*a*) Increases (*b*) Decreases

(*c*) Remain the same

Q. 3. Broilers are generally fed

(*a*) High energy diet (*b*) Low energy diet

(*c*) Very low energy diet

Q. 4. Which Vitamin is necessary for the utilization of Calcium?

(*a*) Vitamin–A (*b*) Vitamin–C

(*c*) Vitamin–D

Q. 5. What mineral is interrelated with iron utilization?

(*a*) Zinc (*b*) Copper

(*c*) Calcium

Q. 6. 'Perosis' is caused due to deficiency of

(*a*) Calcium (*b*) Zinc

(*c*) Manganese (*d*) None of the above

Q. 7. Which Vitamin is a good antioxidant for Vitamin A, carotene and fat?

(*a*) Vitamin D (*b*) Vitamin E

(*c*) Vitamin K (*d*) None of the above

Q. 8. 'Curled-toe paralysis' in chick is caused due to deficiency of

(*a*) Pyridoxine (*b*) Riboflavin

(*c*) Panthothenic acid (*d*) None of the above

Q. 9. 'Crazy chick' diseases is caused due to deficiency of

(*a*) Vitamin- A (*b*) Vitamin-D

(*c*) Vitamin -E (*d*) None of the above

Q. 10. The highest recommended content of Vitamin A, per kg ration by NRC (1989) in breeding chicken is

(*a*) 2,000 I.U. (*b*) 3,000 I.U.

(*c*) 4,000 I.U. (*d*) 5,000 I.U.

Q. 11. The suggested iron content in broiler ration is

(*a*) 20 ppm (*b*) 40 ppm

(*c*) 60 ppm (*d*) 80 ppm

Q. 12. Dietary concentration of choline are determined by the adequacy of Methionine because

(*a*) Methionine can spare choline requirement

(*b*) Methionine can not spare choline requirement

(*c*) Methionine can be converted to choline

(*d*) Methionine has no relation with choline

Q. 13. During the reproductive stage calcium metabolism in birds is about 20 times faster than in mammals and the rate of elimination form the blood is faster by

(*a*) One time (*b*) Three times

(*c*) 5 times (*d*) 20 times

Q. 14. Tryptophan in ration can spare the requirement of

(*a*) Vitamin C

(*b*) Vitamin D_3

(*c*) Vitamin niacin

(*d*) Vitamin cyanocobalamin

Q. 15. Antibiotics are used to the extent of______per tonne of ration in poultry and their feeding should not be continued for more than______days

(*a*) 5–50 mg, 3 days

(*b*) 15–500 mg, 5 days

(*c*) 50–500 mg, 10 days

(*d*) 500–1000 mg, 15 days

Q. 16. Broiler starter practical ration should contain______% protein containing______of lysine,______of leucinie and______% of methionine

(*a*) 20%, 1.00% lysine, 1.25% leucin and 0.75% methionine

(*b*) 19%, 1.75% lysine, 1.75% leucin and 0.95% methionine

(*c*) 22%, 1.50% lysine, 1.00% leucin and 0.50% methionine

(*d*) 23%, 1.22% lysine, 1.54% leucin and 0.83% methionine

Q. 17. Sulphur containing Amino Acids in poultry diets are

(*a*) Methionine, Lysine & Hystidine

(*b*) Methionine, Cystine & Cystein

(*c*) Lysine, Isoleucin & Tryptophan

(*d*) None of the above

Q. 18. End product of protein metabolism in chicken is

(*a*) Ammonia (*b*) Nitrates

(*c*) Urea (*d*) Uric Acid

Q. 19. Limiting Amino Acids in poultry diet are

(*a*) Methionine & Arginine

(*b*) Tryptophan & Hystidine

(*c*) Lysine & Methionine

(*d*) Arginine & Histidine

Q. 20. Precursor of Vitamin 'A' is

(*a*) Retinol (*b*) Carotene

(*c*) Xanthophylls (*d*) Tocopherol

I. F. Multiple Choice Questions (Tick the Correct Answer/s)

Q. 1. Anti-Vitamin K activity is exhibited by

(*a*) Biotin (*b*) Dicumarol

(*c*) Sulphanilamide (*d*) Caproic acid

Q. 2. Gossypol of cotton seed meal reacts with

(*a*) Zinc (*b*) Manganese

(*c*) Iron (*d*) Selenium

Q. 3. 'Nutritional Roup' is due to the deficiency of

(*a*) Vit. A (*b*) Vit. B_6

(*c*) Vit. K (*d*) Vit. E

Q. 4. Pastures are classified into

(*a*) Silage (*b*) Roughage

(*c*) Additives (*d*) Succulent forages

Q. 5. A calorie is the amount of heat required to raise the temperature of 1 g water from

(*a*) 12.5 to 13.5°C (*b*) 14.5 to 15.5°C

(*c*) 15.7 to 16.7°C (*d*) 10.2 to 11.2°C

Q. 6. For determination of Metabolizable energy instrument used is

(*a*) Metabolizable energy meter (*b*) Bomb calorie meter

(*c*) Barometer (*d*) Energy thermometer

Q. 7. Keratins are the protein of

(*a*) Arteries (*b*) DNA

(*c*) Hairs (*d*) Connective tissues

Q. 8. Protamines are basic proteins associated with nucleic acids are rich in

(*a*) Tyrosine (*b*) Tryptophan

(*c*) Methionine (*d*) Arginine

Q. 9. Weight gain per unit weight of protein consumed refers to

(*a*) Biological value

(*b*) Gross protein value

(*c*) Protein Efficiency Ratio (PER)

(*d*) Essential Amino Acid Index

Q. 10. Denaturation of proteins in chicken occurs in

(*a*) Oesophagus

(*b*) Proventriculus and gizzard

(*c*) Crop and pancreas

(*d*) Small and large intestine

Q. 11. Vitamin E (Tocopherol) was discovered by

(*a*) Funk

(*b*) Hopkins

(*c*) Evans and Bishop

(*d*) Mc-Collum and Davis

Q. 12. Maintenance of normal cerebrospinal fluid, pressure is physiological functions of

(*a*) Riboflavin (*b*) Pyridoxine

(*c*) Retinol (*d*) Folic acid

Q. 13. Vitamin H is the old name of

(*a*) Nicotinic acid (*b*) Folic acid

(*c*) Tocopherol (*d*) Biotin

Q. 14. Selenium is an essential component of enzyme

(*a*) Co-enzyme A

(*b*) D-Amino acid oxidase

(*c*) Glutathione per oxidase

(*d*) Cholinesterases

Q.15. "Clubbed down" occur due to deficiency of

(*a*) Pteroyglutamine acid (*b*) Riboflavin

(*c*) Menaquinone (*d*) Cholecalciferol

Q. 16. The vitamin which has cobalt in its structure is

(*a*) B_{12} (*b*) B_2

(*c*) B_1 (*d*) C

Q. 17. Cotton seeds are the chief source of

(*b*) Fat (*c*) Protein

(*d*) Carbohydrate (*e*) Vitamin

Q. 18. Anti-Oxidants are added to feed for

(*a*) Stabilization of Carbohydrates

(*b*) Stabilization of Proteins

(*c*) To prevent Rancidity by stabilization of fats and fat soluble vitamins

(*d*) None of the above

Q. 19. The mineral which prevents the deficiency symptom of Vitamin E is

(*a*) Zinc (*b*) Iron

(*c*) Selenium (*d*) Iodine

Q. 20. The vitamin not essential in the diet of the chicken but needs supplementation during summer stress is

(*a*) Follic acid

(*b*) Pantothenic acid

(*c*) Ascorbic acid

(*d*) Nicotinic acid

Q. 21. Synthesis of carbonic anhydrase enzyme is hindered due to deficiency of

(*a*) Cobalt (*b*) Zinc

(*c*) Iron (*d*) Manganese

Q. 22. Amino acid which is not critical in poultry ration is

(*a*) Lysine (*b*) Arginine

(*c*) Methionine (*d*) Cystine

II. Fill in the Blanks

1. Wrinkled shells occur due to deficiency of_______
2. *Aspergillus flavus* produce toxin known as_______
3. The name of very high energy rich cereal is_______that is used in poultry feeding.
4. The name of higher protein containing oil cake used in poultry feeding_______
5. The vitamins and trace minerals are not premixed together because minerals tend to_______the destruction of the vitamins.
6. The viscosity of molasses is reduced by_______
7. Fat are_______to bring them into liquid.
8. The amount of leaf meal in the poultry mash should_______percent.
9. Antioxidants are added for the_______of fats and fat soluble vitamins.
10. Estrogens are used to improve_______in growing bird.
11. Mustard meal is deficient in amino acid_______
12. Guar meal contain_______inhibitors.
13. Linseed meal is a good source of amino acid_______

14. Yeast is used for vitamin______complex.
15. Maize gluten is the byproduct of the manufacture of______
16. Coconut meal protein is______digested.
17. Sunflower meal is a good source of Arginine and______
18. *Penicillin mycelium* waste is the byproduct obtained from the manufacture of______
19. Cotton seed meal should be used to a______extent due to gossypol content.
20. Molasses can replace cereal up to______percent.
21. The active form of Vitamin D_3 in body is______
22. Absorption of calcium is over______% and that of Mn is______% in highly producing laying birds.
23. The hens daily requirement of calcium for egg shell formation alone, per kg live weight, is about______times that of a high yielding dairy cow.
24. A hen eliminate about______g Calcium during its laying period (250-260 eggs) which is about 25 times that of a diary cow.
25. In layers mean egg weight was found to be increased due to high dietary ME concentration which might be due to higher intake of______
26. Some times cystine is not synthesized in adequate amounts in the birds and hence considered as______
27. Fibrous proteins contains the______which are the main proteins of______
28. The chemical name of Vitamin D_2 is______where as D_3 is______
29. ______concerned with night vision is a______pigment.

30. The enzyme like______break down fat into______ and ______

31. The inorganic element present in the Arginase is______It splits Arginine into______and______

III. State True or False for the Following Statements

1. Protein requirement of breeding quail is 24 percent.
2. Marble grit is required to supplement calcium and phosphorus.
3. Cannibalism occurs due to sodium deficiency in poultry mash.
4. Units are used to express Vitamin A and D because different forms of vitamin have different biological activities.
5. Energy requirements of hens at different ambient temperature may vary substantially.
6. The amount of calcium present in limestone is about 38 percent.
7. Low protein diet may not retard the growth of pullet.
8. Poultry do not require abundant energy to maintain temperature.
9. Protein and energy relationship may affect the size of egg.
10. Poultry require high level of crude fiber.
11. Continuous feeding of the cotton seed meal results in Mottling of Egg yolk.
12. Curled toe paralysis in chickens is due to deficiency of Vitamin 'A' in the diet.
13. Feather meal is a rich source of fat.

14. Encephalomalacia is caused due to deficiency of Vitamin B_6.

15. Broiler starter ration should contain higher ME concentration in relation to finisher.

16. For breeding males less amount of protein is desirable in the ration for better performance.

17. Methionine cannot be converted to cystine during the metabolism.

18. In growing chicken higher is the energy level in the diet fatter will be the birds.

19. In growing young turkeys the growth rate is strongly influenced by the dietary average energy levels.

20. Feed containing higher energy level show better feed efficiency in turkeys.

21. Calcium and zinc can form triple complexes with phytic acid. Such phytates can be broken down by the strong activity of phytase enzyme present in birds in comparison to ruminants.

22. High dietary levels of vitamin A or E have been found to enhance the immune responses in chicks.

23. In broiler breeders the concentration of biotin in yolk increases with age and those of pyridoxine and cobalamin decrease.

24. In young guinea fowl energy level above 11.8 MJ ME/kg did not indicate further growth response. However, addition of lipids in diet changes chemical composition of gain.

25. Proteins are complex organic compounds of low molecular weight.

26. Casein of milk contains phosphoric acid as prosthetic group.

27. Hay, straw, seed hulls are classified as succulent forages.
28. Tyrosine is synthesized from proline.
29. Certain amino acids can be produced from others by a process known as transamination.
30. In the retina, rods are concerned with vision in bright light.
31. Vitamin B_{12} is widely distributed in foods of sea origin.
32. Avidin binds with biotin and inactivates it.
33. Pyrithiamine is an anti-metabolite of pyridoxines.
34. Sodium appears to be almost completely absent from blood cells.
35. Polyneuritis in chicken can occur when they are fed unpolished rice.
36. Metabolic Faecal Nitrogen and Endogenous Urinary Nitrogen are taken into consideration while calculating Gross Protein Value.
37. Histidine is an essential amino acid for poultry.
38. Magnesium is present in both erythrocytes and plasma.
39. Cod liver oil is an excellent source of Vitamin A.

IV. Match the Following

1.	Gumbaro	(*a*)	Rupture of liver
2.	Fatty liver	(*b*)	Lysine rich
3.	Vitamin D_3	(*c*)	Curled toe paralysis
4.	Soya	(*d*)	Tremors on head of chick
5.	Manganese	(*e*)	Inflamed bursa
6.	Riboflavin	(*f*)	Soft bones
7.	MDV	(*g*)	Perosis

8.	AE	(*h*)	Pecking
9.	Salt	(*i*)	Rancidity
10.	Antioxidant	(*j*)	HVT
11.	B (1–4) glycosidic linkage	(*k*)	Proteins
12.	Hydroxy lysine	(*l*)	Hemicellulose
13.	Carnitine	(*m*)	PUFA
14.	Amprolium	(*n*)	Cellulose
15.	Chlortetracycline potentiation	(*o*)	Fatty acid carrier
16.	Quaternary structure	(*p*)	Biotin
17.	Pentosans	(*q*)	Trimethoprim
18.	Linoleic acid	(*r*)	Sodium sulphate
19.	Dermatitis	(*s*)	Collagen
20.	Sulpha potentiation	(*t*)	Coccidiosis

Answers

I. A. Multiple Choice Questions

1	(*c*)	5	(*b*)	9	(*c*)	13	(*b*)
2	(*b*)	6	(*a*)	10	(*c*)	14	(*d*)
3	(*d*)	7	(*b*)	11	(*a*)		
4	(*d*)	8	(*b*)	12	(*c*)		

I. B. Multiple Choice Questions

1	(*c*)	6	(*a*)	11	(*c*)	16	(*b*)
2	(*c*)	7	(*c*)	12	(*b*)	17	(*d*)
3	(*b*)	8	(*c*)	13	(*a*)	18	(*c*)
4	(*c*)	9	(*a*)	14	(*d*)	19	(*d*)
5	(*c*)	10	(*a*)	15	(*a*)	20	(*d*)

I. C. Multiple Choice Questions

1	(*b*)	6	(*a*)	11	(*b*)	16	(*c*)
2	(*c*)	7	(*c*)	12	(*a*)	17	(*a*)
3	(*d*)	8	(*a*)	13	(*d*)	18	(*a*)
4	(*b*)	9	(*d*)	14	(*c*)	19	(*a*)
5	(*a*)	10	(*d*)	15	(*a*)	20	(*c*)

I. D. Multiple Choice Questions

1	(*b*)	6	(*c*)	11	(*a*)	16	(*a*)
2	(*b*)	7	(*d*)	12	(*c*)	17	(*d*)
3	(*a*)	8	(*d*)	13	(*a*)	18	(*c*)
4	(*c*)	9	(*a*)	14	(*c*)	19	(*d*)
5	(*c*)	10	(*d*)	15	(*d*)	20	(*c*)

I. E. Multiple Choice Questions

1	(*b*)	6	(*c*)	11	(*d*)	16	(*d*)
2	(*b*)	7	(*b*)	12	(*a*)	17	(*b*)
3	(*a*)	8	(*b*)	13	(*c*)	18	(*d*)
4	(*c*)	9	(*c*)	14	(*c*)	19	(*c*)
5	(*b*)	10	(*c*)	15	(*c*)	20	(*b*)

I. F. Multiple Choice Questions

1	(*b*)	7	(*c*)	13	(*d*)	19	(*c*)
2	(*c*)	8	(*d*)	14	(*c*)	20	(*c*)
3	(*a*)	9	(*c*)	15	(*b*)	21	(*b*)
4	(*d*)	10	(*b*)	16	(*a*)	22	(*b*)
5	(*b*)	11	(*c*)	17	(*b*)		
6	(*b*)	12	(*c*)	18	(*c*)		

Answers for Q. No. II Fill in the Blanks

1. Copper
2. Aflatoxin
3. Maize
4. Groundnut cake
5. Accelerate

6. **Heating**
7. **Heated**
8. **5**
9. **stabilization**
10. **growth**
11. **lysine**
12. **tripsin**
13. **tryptophan**
14. **B**
15. **Starch**
16. **Poorly**
17. **Methionine**
18. **Penicillin**
19. **Limited**
20. **5 to 10**
21. **1, 25 dihydroxy cholecalciferol**
22. **60% for Ca, 2-5% for Mn**
23. **25**
24. **500**
25. **ME**
26. **EAA**
27. **Collagens, connective tissues.**
28. **Ergocalciferol, cholecalciferol**
29. **Rhodopsin, bright red**
30. **Lipase, fatty acids and glycerols**
31. **Magnesium, ornithine and urea.**

III. True or False

1	True	11	True	21	False	31	True
2	False	12	False	22	True	32	True
3	True	13	False	23	True	33	False
4	True	14	False	24	True	34	True
5	True	15	False	25	False	35	False
6	True	16	True	26	True	36	False
7	False	17	False	27	False	37	True
8	False	18	True	28	False	38	True
9	True	19	False	29	True	39	True
10	False	20	True	30	False		

IV. Match the Following

1	(*e*)	6	(*c*)	11	(*n*)	16	(*k*)
2	(*a*)	7	(*j*)	12	(*s*)	17	(*l*)
3	(*f*)	8	(*d*)	13	(*o*)	18	(*m*)
4	(*b*)	9	(*h*)	14	(*t*)	19	(*p*)
5	(*g*)	10	(*i*)	15	(*r*)	20	(*q*)

Chapter 14

Poultry Health Care

I. Multiple Choice Questions
(Pick the Write Choice/s)

Q. 1. What is the incubation period of New Castle disease (Ranikhet Disease)?

(*a*) Fortnight (*b*) One week

(*c*) 2-4 days (*d*) One month

Q. 2. What is the nature of Mukteshwar strain vaccine?

(*a*) Attenuated virus chick embryo vaccine

(*b*) Dead virus chick embryo vaccine

(*c*) Live virus chick embryo vaccine

(*d*) None of the above

Q. 3. Avian encephalomyelitis is transmitted

(*a*) Through air (*b*) Through feed

(*c*) Through water (*d*) Through egg

Q. 4. Which of the following diseases is caused by *Pasteurella multocida*

(*a*) Fowl typhoid (*b*) Fowl cholera

(*c*) Bacillary white diarrhoea (*d*) None of the above

Q. 5. Which of the following drugs is most effective in treating or prevention of CRD?

(*a*) Penicillin (*b*) Tiamutin

(*c*) Streptomycin (*d*) Tetracycline

Q. 6. Chicks of which age are most susceptible to brooder pneumonia

(*a*) Six week old (*b*) 4 week old

(*c*) 2 week old (*d*) Chicks of few days

Q. 7. Chicken are vaccinated against Marek's disease

(*a*) Immediately after hatch

(*b*) On the 4th day after hatch

(*c*) On the 21st day after hatch

(*d*) All the above

Q. 8. Petichial hemorrhages in the proventriculus is an indication of

(*a*) IBD (*b*) Coccidiosis

(*c*) R.D. (*d*) Marek's diseases

Q. 9. Most of the basic breeders and hatchery men test their flock for

(*a*) Salmonellosis (*b*) Mycoplasmosis

(*c*) All the above (*d*) None of the above

Q. 10. Perosis is usually caused due to deficiency of

(*a*) Iodine (*b*) Manganese

(*c*) Zinc (*d*) All the above

Q. 11. Vertical transmission refers to spread of disease through

(*a*) Ovary (*b*) Uterus

(*c*) Embryo (*d*) All the above

Q. 12. The most potent method of control of lymphoid Leukosis is

(*a*) By vaccination (*b*) By use of drugs

(*c*) Combination of (*a*) & (*b*) (*d*) None of the above

Q. 13. Ranikhet disease is also known as

(*a*) New Castle disease

(*b*) Pseudo fowl pest/ Pseudo fowl plague

(*c*) Avian Diarrhoea

(*d*) All the above

Q. 14. Characteristic symptom of Infectious Bronchitis in chicken is

(*a*) Reduced egg production

(*b*) Soft/Rough Shelled and Mishapen eggs

(*c*) Watery albumen or Yolk sac separation

(*d*) All the above

Q. 15. To differentiate Infectious bronchitis, Infectious Laryngotracheatis, and Chronic respiratory disease (Mycoplasma) upon embryo inoculation following observation could be made for respective diseases

(*a*) Curling and dwarfing of embryos in Infectious bronchitis

(*b*) Necrotic pock lesions in I. L.T.

(*c*) Death of embryo due to affected yolk sac

(*d*) All of the above

Q. 16. Following are the neoplastic diseases in poultry

(*a*) Marek's disease and Avian lymphoid leucosis

(*b*) Reticulendothelisis

(*c*) Lympho Proliferative disease

(*d*) All the above

Q. 17. Following are the synonyms of Inclusion Body Hepatitis (I.B.H.) disease in Poultry

(*a*) Hydropericardium syndrome

(*b*) Leechi/ Angora disease

(*c*) Egg drop syndrome -76 (E.D.S.-76)

(*d*) CELO-Chicken embryo lethal orphan

(*e*) All the above

Q. 18. Reoviral infections in chicks cause

(*a*) Malabsorption syndrome (Stunted chick syndrome),

(*b*) Viral arthritis (Tinosynovitis),

(*c*) Proventriculitis, pancreatic enlargement

(*d*) Helicopter disease

(*e*) All the above

II. Fill in the Blanks

1. Two forms of fowl pox observed in poultry are Cuteneous form and_______form

2. Avian lymphoid leucosis is also known as_______.

3. The term colibacillosis includes infections such as_______,_______,_______,and_______caused by *E. coli* organism.

4. Sever coughing and expectoration of mucus and blood stained exudates and blood clot followed by death of birds in 1 to 3 days is indication of_______disease.

5. Petechial haemorrhages in the leg and thigh muscle and presence of enlarged, inflamed, oedematous and cream coloured bursa is the indication of the_______disease.

6. Caecal coccidiosis is caused by_______spp.

7. _______and_______are the most prevalent species of protozoa causing coccidiosis in poultry.

8. _______,_______and_______spp. causes most pathogenic coccidiosis in poultry.

9. Spirochaetosis in chicken is caused by_______organism that is transmitted by a soft tick called_______.

10. In adult birds, the characteristic lesion of an abnormal ovary with ova irregular, cystic misshapen discoloured and pedunculated with prominent thickened stalk is indication of_______disease.

III. State True or False for Below Mentioned Statements

1. Lasota, F and B_1 strains of Ranikhet disease virus belongs to mesogenic pathotype.

2. R_2B, Roakin, and Kumrao are lentogeneic strains of Ranikeht disease.

3. Mukteshwar, Hearts, G. B., Melano and Texas are strains of Velogenic phathotype of Ranikhet disease.

4. Infectious bronchitis is caused by Corona virus which has 12 strains and most important one are *Baudette*, *Connecticut*, and *Massachusetts* strains.

5. Vaccination of chickens for Infectious Laryngotracheitis disease is not banned in India.

6. Perivascular cuffing of lymphocytes in central nervous system of chicken is the pathognomonic lesion of Ranikhet disease.

7. Three major pathotypes of Ranikhet disease are lentogenic, mesogenic and velogenic types.

8. Two types of Velogenic forms of New Castle disease are Velogenic Viscerotropic N. D. (Doyel's form) and Velogenic Neurotrophic N.D. (Beache's form).

9. Infectious bursal disease (I.B.D./Gumbaro disease) is caused by double stranded DNA Reo virus of Birna viridae family.

10. Fowl pox and chicken pox are one another same and they are caused by same virus.

11. The term REO in Reo virus stands for Respiratory Enteric Orphan virus.

12. The term MATSA stands for MD Tumor Associated Surface Antigen which is useful for confirmative diagnosis of Marek's disease in poultry.

13. The term COFAL stands for complement fixation test for avian leucosis complex.

14. Avian Encephalitis also known as 'epidemic tremor' results in nervous symptoms in chicken is caused by virus belonging to Picorna viridae family.

15. Avian influenza also known as 'Fowl plague' is caused by the virus belonging to orthomyxo viridae family.

16. Chicken infectious anemia agent is caused by Circodena virus.

17. H5N1 strain is the highly pathogenic and highly zoonotic strain of avian influenza virus.

18. Avian influenza is caused by the enveloped RNA virus belonging to Orthomyxo viridae that are further subgrouped based on Haemaglutinin (H) and Neuraminidase (N) antigens.

19. At present there are 13 recognized H subtypes and 9 N subtypes of avian influenza virus and most of the 117 possible combinations have been isolated from avian species.

20. Finding of dark red or black swollen liver with a distinctive coppery bronze sheen is and indication of Fowl Typhoid infection in chicken.

21. The rapid serum agglutination test (RSA)/Tube agglutination test/ Haemagglutination inhibition test are among the most commonly used tests for diagnosis of Mycoplasma infections in poultry.

22. Marek's disease affects chicks of less than 6 weeks old while lymphoid leucosis affects birds of 16 weeks old or above.

23. Enveloped RNA virus (Avian Paramyxovirus type-1–PMV-1) belonging to paramyxoviridae family causes Ranikhet/ New Castle disease in poultry.

24. The chick anaemia agent is also known as Blue wing disease.

25. Infectious avian encephalomyelitis is also known as 'epidemic tremor'

IV. A. Match the Following Diseases with their Respective Causative Agents

1.	Bacillary white diarrhoea	(*a*)	*Bordetella bronchoseptica / avisepticum*
2.	Fowl Typhoid	(*b*)	*Clostridium colinum*
3.	Paratyphoid	(*c*)	*Salmonella* spp. other than *pullorum* & *gallinarum*
4.	Fowl Cholera	(*d*)	*Campalobacter jejuni*
5.	Infectious Coryza /Fowl coryza	(*e*)	*Mycoplasma synoviae*
6.	CRD	(*f*)	*Cl. Botulinum*
7.	Brooder pneumonia	(*g*)	*E. coli*
8.	Favus (White comb)	(*h*)	*Candidia albicans*
9.	Moniliasis (Candidiasis/ Thrush)	(*i*)	*Trichophyton magnini*
10.	Ornithosis	(*j*)	*Chlamidia albicans*
11.	Hjarres disease (Coligranuloma)	(*k*)	*Aspergillus fumigatus*
12.	Limber neck	(*l*)	*Mycoplasma galli-septicum*
13.	Infectious synovitis	(*k*)	*Haemophilus galli-narum*
14.	Avian infectious/ Vibrionic hepatitis	(*m*)	*Pasteurella multoceda*
15.	Quil disease (Ulcerative enteritis)	(*n*)	*S. gallinarum*

16.	Bacterial endocarditis	(*o*)	*S. pullorum*
17.	Gangrenous dermatitis	(*p*)	*Cl. Perfringens type* C
18.	Necrotic enteritis	(*q*)	*Cl. Perfringens type* A/ *Cl. Septicum*

IV. B. Match the Following Etiological Agents with their Culture Media

1.	*Mycobacterium tuberculosis*	(*a*)	P.P.L.O. agar/Broth/ Fray's media
2.	*Asperigillus fumigatus*	(*b*)	Brilliant green agar media
3.	*Clostridium botulinum*	(*c*)	Robert's cooked meat media
4.	*Salmonella pullorum*	(*d*)	Sabroud's media
5.	Mycoplasma	(*e*)	Dorset's egg media

IV. C. Match the Following Common Names of Parasites with their Scientific Names

1	Wing louse	(*a*)	*Knemedocoptes mutans*
2	Shaft louse	(b)	*Knemedocoptes gallinae*
3	Chicken head louse	(*c*)	*Dermanyssus gallinae*
4	Fowl tick	(*d*)	*Argas parsicus*
5	Red mite	(*e*)	*Cuelotogaster hetero-grapher*
6	Feather mite	(*f*)	*Genocotes gallinae*
7	Scaly leg mite	(*g*)	*Lipeurus caponis*
8	Gape worm	(*h*)	*Heterakis gallinarum*
9	Smallest nematode in poultry	(*i*)	*Ascaridia* spp.
10	Largest nematode in poultry	(*j*)	*Capillaria* spp.
11	Caecal nematode (typhilitis)	(*k*)	*Syngamus trachea*

Answers

I. Multiple Choice Questions

1	(*c*)	6	(*d*)	11	(*d*)	16	(*d*)
2	(*c*)	7	(*a*)	12	(*d*)	17	(*e*)
3	(*d*)	8	(*c*)	13	(*d*)	18	(*e*)
4	(*b*)	9	(*c*)	14	(*d*)		
5	(*b*)	10	(*b*)	15	(*d*)		

II. Fill in the Blanks

1. Diphtheritic form
2. Big liver disease
3. Yolk sac infection, Coligranuloma, Egg peritonitis and Colisepticaemia.
4. Infectious Laryngotracheitis
5. Infectious Bursal disease.
6. *Eimeria tenella*
7. *Eimeria acervulina, Eimeria maxima*
8. *Eimeria tenella, Eimeria acervulina and Eimeria maxima*
9. *Borrelia anserina, Argas persicus*
10. Pullorum disease

III. True or False

1	False	8	True	15	True	22	True
2	False	9	True	16	True	23	True
3	True	10	False	17	False (low zoonotic)	24	True
4	True	11	True	18	True	25	True
5	False	12	True	19	True		
6	True	13	True	20	True		
7	True	14	True	21	True		

IV. A. Match the Following

1	(*o*)	6	(*l*)	11	(*g*)	16	(*a*)
2	(*n*)	7	(*k*)	12	(*f*)	17	(*q*)
3	(*c*)	8	(*i*)	13	(*e*)	18	(*p*)
4	(*m*)	9	(*h*)	14	(*d*)		
5	(*k*)	10	(*j*)	15	(*b*)		

IV. B. Match the Following

1	(*e*)	3	(*c*)	4	(*b*)	5	(*a*)
2	(*d*)						

IV. C. Match the Following

1	(*g*)	4	(*d*)	7	(*a*)	10	(*i*)
2	(*f*)	5	(*c*)	8	(*k*)	11	(*h*)
3	(*e*)	6	(*b*)	9	(*j*)		

Chapter 15

Post Harvest Processing

I. Multiple Choice Questions (Tick Correct Answer/s)

Q. 1. Yolk constitutes around

(*a*) 43-45 per cent of whole egg

(*b*) 55-57 per cent of whole egg

(*c*) 22-24 per cent of whole egg

(*d*) 30-32 per cent of whole egg

Q. 2. A hen's egg contains around

(*a*) 10-11 grams of protein

(*b*) 12-13 grams of protein

(*c*) 6-7 grams of protein

(*d*) 14-15 grams of protein

Q. 3. A component of egg white having antibacterial activity is known as

(*a*) Avidin (*b*) Conalbumin

(*c*) Albumen (*d*) Lysozyme

Q. 4. Minimum requirements for pasteurization of whole egg blends (less than 2% added non- egg ingredients) are

(*a*) 142°F for 3.5 min (*b*) 134°F for 3.5 min

(*c*) 146°F for 3.5 min (*d*) 132°F for 3.5 min

Q. 5. Most tender meat can be obtained from

(*a*) Turkey (*b*) Geese

(*c*) Broiler chickens (*d*) Duck

Q. 6. Compared to red meats poultry meats contains a higher proportion of

(*a*) Saturated fatty acids

(*b*) Unsaturated fatty acids

(*c*) Both (*a*) and (*b*)

(*d*) None of the above

7. Freeze dried poultry products are generally stored at

(*a*) Ambient temperature

(*b*) Refrigerated temperature

(*c*) Freezing temperature

(*d*) None of the above

Q. 8. By irradiation oxidative deterioration of poultry meat is

(*a*) Retarded

(*b*) Hastened

(*c*) Not changed

(*d*) None of the above

Q. 9. Chicken pickle and tandoori chicken are foods of

(*a*) Indigenous origin (*b*) Exotic origin

(*c*) Both (*a*) and (*b*) (*d*) None of the above

Q. 10. Warner Bratzler shear press is used to measure

(*a*) Flavor (*b*) Colour

(*c*) Tenderness of poultry meat (*d*) None of the above

Q. 11. As an egg ages and loses weight, the specific gravity:

(*a*) Increases (*b*) Decreases

(*c*) Remain unchanged

Q. 12. Conalbumin is responsible for chelation of

(*a*) Iron, Zinc, Cobalt (*b*) Iron, Cobalt, Silver

(*c*) Iron, Zinc, Copper

Q. 13. In baked products, eggs when added acts as

(*a*) Thickening agent (*b*) Clarifying agent

(*c*) Leavening agent

Q. 14. Eggs are primarily pasteurized to destroy

(*a*) Salmonella

(*b*) Molds and yeast cells

(*c*) Fecal coliforms

Q. 15. Efficiency of pasteurization of egg yolk can be detected by

(*a*) Catalase test (*b*) Phosphatase test

(*c*) Alpha amylase test

Q. 16. Egg white or whole egg used for preparation of egg powder is fermented before drying to remove

(*a*) Glucose

(*b*) To develop desired acidity in the finished dried product

(*c*) To prevent enzymatic browning

Q. 17. In general spoilage of eggs is caused more by

(*a*) Bacteria (*b*) Yeast

(*c*) Molds and yeast

Q. 18. Bleeding of chickens is more perfect

(*a*) If the birds are fasted over night with free access to potable water

(*b*) Well fed before bleeding

(*c*) No water is given

Q. 19. Meat from imperfectly bled chicken will have

(*a*) Better shelf life

(*b*) Poor shelf life

(*c*) Shelf life will not be affected

Q. 20. The formation of dark greenish discolouration on the surface of egg yolk of hard boiled eggs is due to

(*a*) The H_2S combines with the iron of the yolk giving a dark green colour

(*b*) Due to Maillard reaction

(*c*) Due to presence of penicillium molds

Q. 21. Chicken egg is a poor source of

(*a*) Vitamin 'A' (*b*) Vitamin 'C'

(*c*) Vitamin 'D' (*d*) Vitamin 'E'

Q. 22. The fatty acid which is not synthesized in the body of chicken is

(*a*) Linoleic acid (*b*) Linolenic acid

(*c*) Oleic acid (*d*) Lauric acid

Q. 23. Predominant spoilage bacteria of egg is

(*a*) *Salmonella* (*b*) *E. coli*

(*c*) *Pseudomonas* (*d*) *Campylobacter*

II. Fill in the Blanks

1. The cholesterol content of a standard egg is around______

2. The total time interval between the releasing of the yolk from the ovary and laying of the egg is usually ______hours.

3. Eggs consist of around______% shell.

4. Per capita availability of eggs per annum in India is______.

5. Calorie value of an egg is______
6. "Egg nog" is the product prepared from eggs and______
7. There are around______to______pores in chicken eggs shell.
8. ______is a transparent protein coating in the surface of the shell.
9. Eggs consist of______% yolk.
10. The egg shell consists of around______% Calcium carbonate.
11. The thick layer immediately surrounding the yolk is called the inner thick or______layer.
12. A hen egg contains around______grams of protein.
13. A component of egg white having antibacterial activity is known as______
14. As an egg loses its weight and the specific gravity______
15. Eggs consist of around______% albumin
16. Fertilization of an egg takes place in the______
17. The average weight of an quail egg is______
18. Conalbumin is responsible for chelation of______, ______and______
19. In baked products eggs when added acts as______
20. Efficiency of pasteurization of egg yolk can be detected by ______
21. In general spoilage of eggs is caused more by (bacteria/ yeast/ mould).
22. The most abundant protein in egg white is______

23. Eggs contain generous quantities of all the essential vitamins except______

24. The egg yolk colour is unrelated to nutritional value, it only reflects the amount of______present in feed.

25. Minimum requirement for pasteurization of whole egg blends are______

26. Peeling of shell of a freshly laid boiled egg is (easier/difficult) than an aged boiled egg.

27. During storage, the (acidity/alkalinity) of the egg white increases.

28. On freezing and storing raw egg below–21°C its functional properties are (altered/not altered).

29. Gelation of egg yolk can be avoided by the incorporation of______% of sugar before freezing.

30. Egg white or whole egg used for preparation of egg powder is fermented before drying to remove______

31. The thickness of chicken egg shell varies from______to ______mm.

32. Eggs are washed in 2% of______Solution at 40 °C for 3 min before making a product.

33. As per BIS standard the *Salmonella* content in egg powder should be______

34. In Mayonnaise and salad dressing egg yolk is used as______agent.

35. The thickness of outer and inner shell membranes range between______and______mm.

36. The outer and inner shell membranes of an egg constitutes about______% of the egg weight.

37. The egg shell composed of______% Calcium carbonate, ______% Calcium phosphate,______% Magnesium carbonate and______% Organic matter.

38. The height of air shell in a fresh chicken egg is around______

39. The protein and fat content of entire egg is______% and ______% respectively.

40. The standard weight of an commercial chicken egg is______

41. Shell membranes are added to the egg in______during egg formation.

42. Fertilization of an egg takes place in the______

43. The pH of yolk in freshly laid egg is generally about______

44. Thick white and practically all egg white protein secreted from______during egg formation.

45. There are around______pores in a fresh chicken egg shell.

46. Eggs are washed in 2% of sodium hypochlorite solution at______°C for 3 min before making a product.

47. Good quality egg can be kept in cold store at 0 °C to 1.7 °C and a relative humidity of 70–80% for about______months.

48. Breast forms______% of the eviscerated portion of chicken.

49. The eviscerated percent of broilers at market weight is______

50. The egg spoilage is caused by physicochemical and______

51. Eggs are pasteurized primarily to destroy bacteria pathogenic to humans, especially______spp.

52. Giblets include______, ______ and ______.

53. _______appear to be the most active antioxidant compounds and_______the principal bactericidal agent in the smoke used for smoking poultry.

54. "Modified Kosher" killed birds are those where the_______vein is severed just below the ear so that the windpipe and_______remain uncut.

III. Write True or False Against Each Statement

1. Cuticle (bloom) on the surface of egg shell helps to prolong the shelf life of an egg.
2. Germinal disc is present only in fertilized eggs.
3. Bloodspots inside an egg are caused by rupture of small blood vessels of developing embryo.
4. The colour of yolk is affected by the pigments present in the diet of hen.
5. The protein content of egg white is more than egg yolk.
6. Fat content in egg yolk is higher than egg white.
7. Egg is recommended to children's as it is a rich source of Vitamin C and Calcium.
8. Egg white contains exceptionally high level of cholesterol.
9. As the egg ages the size of the air space decreases.
10. When the egg is broken onto a flat dish and if white is thin and watery and yolk flattens out or breaks–it is an indication of freshness of an egg.
11. The egg shell consists of around 94-97 percent calcium carbonate.
12. The germinal disc on the surface of the egg yolk is always present, whether the egg is fertilized or not.

13. The egg yolk colour is unrelated to nutritional value; it only reflects the amount of xanthophylls pigment in the hen's diet.

14. Albumen contains more lipids and less moisture than yolk.

15. Egg protein is one of the highest quality proteins known for human food.

16. Eggs contain generous quantities of all the essential vitamins including Vitamin C.

17. Egg yolk contains natural emulsifiers.

18. Pink albumin can be caused by cottonseed and soybean meal in hen's diet.

19. In general, shrinkage is greater in older or heavier birds than in young birds and light weight ones.

20. In ducks, among the cut-up parts, legs and thighs produce highest cut percentage based on carcass weight.

21. Nutrient losses of poultry meat from freezing and thawing are primarily due to "drip".

22. In poultry, white meat contains less protein than dark meat.

23. Salt in the cure influences the flavour and texture of meat and acts as a preservative.

24. Canned chicken usually contains a chicken with chicken broth but without giblets.

25. The faster the poultry carcasses pass through the zone of maximum crystal formation, the larger crystals in the frozen products.

26. Enzyme papain is commonly used for increasing the self life of poultry meat.

27. Cloudiness in eggs indicates reject eggs unfit for human consumption.

28. Faster freezing time helps the frozen carcass to remain white.

29. Shape index is obtained by dividing length of the egg with width of the egg and multiplying by 100.

30. Broiler contains about 151 calories per 100 g of meat.

31. The salvage value of spent hen is less than 10%.

32. The biological value of egg protein and meat protein is nearly same.

IV. Tick Mark the Correct Answer

1. Peeling of shell of a freshly laid boiled egg is easier/ difficult than an aged boiled egg.

2. During storage, the acidity/ alkalinity of the egg white increases.

3. On freezing & storing raw egg yolk below –21°C its functional properties are altered/ not altered.

4. Gelation of egg yolk can be avoided/ cannot be avoided by the incorporation of 5% or more of sugar before freezing.

5. Good quality egg can/ cannot be kept in cold store at 0° to –1.7°C and relative humidity of 70-80% for about six months.

6. Poultry should be frozen slow/ fast to retain most of the natural bloom of a freshly dressed chicken.

7. Formaldehyde / methyl alcohol present in wood smoke is the principle bactericidal agent which helps to preserve smoked poultry.

8. Vinegar, or acetic acid used in poultry pickle is a Class II/ Class I preservative.

9. The legal level permitted for nitrite in finished poultry products is 500 ppm / 155 ppm at the time of formulation.

10. The legal limits for added residual phosphates in finished meat products including poultry is 0.5/ 1.5%.

11. Ionizing radiation has been allowed/ not allowed by the Government of India for bulk preservation of poultry for export purposes.

12. Scalding of chicken is usually done at 58–60°C for 1-2 min/ 78-90°C for 1-2 min.

13. The moisture content in freeze dried raw poultry meat is about 1.5% / 7.5%.

14. Canned chicken can/ cannot be stored at room temperature.

15. Yolk colour can be/ cannot be determined by using Roche colour fan.

V. A. Match the Following

1.	Olive coloured yolks	(*a*)	Shell thickness
2.	Yolk colour	(*b*)	Carcinogenic compounds
3.	Monosodium Glutamate	(*c*)	Adhesive
4.	Lysozyme	(*d*)	Xanthophyll
5.	Avidin	(*e*)	Cotton seed meal
6.	Egg powder	(*f*)	Irradiation
7.	Technical Albumen	(*g*)	Lyses some bacteria
8.	Paper thickness gauge	(*h*)	Bind biotin
9.	CO 60	(*i*)	Flavour enhancer
10.	Nitrosoamines	(*j*)	Spray drying

V. B. Match the Following

1.	Haugh Unit	(*a*)	Shell porosity
2.	Weight loss of egg	(*b*)	Avidin
3.	Dried albumen for non food purpose	(*c*)	Canned product
4.	Conabulmin	(*d*)	Low keeping quality
5.	Combination with biotin and its inactivation	(*e*)	Albumen quality
6.	Egg grading	(*f*)	Nitrosohaemo-chrome
7.	Bone darkening	(*g*)	Tandoori chicken
8.	Curing	(*h*)	Technical albumen
9.	Poor bleeding of poultry	(*i*)	Yolk quality
10.	Chicken-a-la king	(*j*)	Chelation of iron, zinc and copper
		(*k*)	Frozen poultry
		(*l*)	Chilling
		(*m*)	Candling
		(*n*)	Smoking

Answers

I. Multiple Choice Questions

1	(*d*)	7	(*a*)	13	(*c*)	19	(*b*)
2	(*c*)	8	(*b*)	14	(*a*)	20	(*a*)
3	(*d*)	9	(*a*)	15	(*c*)	21	(*b*)
4	(*a*)	10	(*c*)	16	(*a*)	22	(*a*)
5	(*c*)	11	(*b*)	17	(*a*)	23	(*c*)
6	(*b*)	12	(*c*)	18	(*a*)		

II. Fill in the Blanks

1. 210–250 mgs.
2. 25 hours
3. 11%
4. 40–45 eggs
5. 100 calorie
6. Milk
7. 1700 to 7500
8. Cuticle
9. 30–32%
10. 94–97%
11. Chalaziferous layer
12. 6–7 gm of protein
13. Lysozyme

14. Decreases
15. 58-60
16. Infundibulum
17. 10 gm
18. Iron, Zinc and Copper
19. Leavening agent
20. Alpha amylase test
21. Bacteria
22. Ovoglobulin
23. Vitamin C
24. Xanthophyll
25. 142 °F for 3.5 mins.
26. Difficult
27. Alkalinity
28. Altered
29. 5%
30. Sugar
31. 0.28 to 0.38 mm
32. Sodium Hypochlorite solution
33. Zero
34. Emulsifying agent
35. 0.01–0.02 mm
36. 0.6%
37. 94, 1, 1, 4
38. 3.5 mm
39. 13 and 11
40. 56 gm
41. Uterus.
42. Infundibulum
43. 6

44. Magnum
45. 1700 to 7500
46. 40 °C
47. Six months.
48. About 50%
49. 75%
50. Microbiological
51. *Salmonella* spp.
52. Heart, Liver and Gizzard
53. Phenols and Formaldehyde
54. Jugular and Esophagus

III. Write True or False

1	True	9	False	17	True	25	False
2	False	10	False	18	True	26	False
3	False	11	True	19	False	27	False
4	True	12	True	20	False	28	True
5	False	13	True	21	True	29	False
6	True	14	False	22	False	30	True
7	False	15	True	23	True	31	False
8	False	16	False	24	True	32	False

IV. Tick Mark the Correct Answer

1. Difficult
2. Alkalinity
3. Altered
4. Avoided
5. Can
6. Fast

7. Formaldehyde
8. Class I
9. 155 ppm
10. 0.5%
11. Not allowed
12. 58-60°C
13. 1.5%
14. Can
15. Can be

V. A. Match the Following

1.	(*e*)	4.	(*g*)	7.	(*c*)	10.	(*b*)
2.	(*d*)	5.	(*h*)	8.	(*a*)		
3.	(*i*)	6.	(*j*)	9.	(*f*)		

V. B. Match the Following

1.	(*e*)	4.	(*j*)	7.	(*k*)	10.	(*c*)
2.	(*a*)	5.	(*b*)	8.	(*f*)		
3.	(*h*)	6.	(*m*)	9.	(*d*)		

Chapter 16

Alternate Poultry Species

I. Multiple Choice Questions (Tick Correct Option/s)

Q. 1. **The chromosome number of Guinea fowl is**

(*a*) 78 (*b*) 80

(*c*) 82 (*d*) 86

Q. 2. **The chromosome number of Turkey is**

(*a*) 78 (*b*) 80

(*c*) 82 (*d*) 86

Q. 3 **The chromosome number of Duck is**

(*a*) 78 (*b*) 80

(*c*) 82 (*d*) 86

Q. 4. **The chromosome number of Muscovy Duck is**

(*a*) 78 (*b*) 80

(*c*) 82 (*d*) 86

Q. 5. The chromosome number of Pigeon is

(*a*) 78 (*b*) 80

(*c*) 82 (*d*) 86

Q. 6. The chromosome number of Pheasants is

(*a*) 78 (*b*) 80

(*c*) 82 (*d*) 86

Q. 7. The chromosome number of Japanese quail is

(*a*) 78 (*b*) 80

(*c*) 82 (*d*) 86

Q. 8. The chromosome number of Ostrich is

(*a*) 78 (*b*) 80

(*c*) 82 (*d*) 86

Q. 9. The chromosome number of Emus is

(*a*) 78 (*b*) 80

(*c*) 82 (*d*) 86

Q. 10. The ancestor of all domestic duck breeds

(*a*) Wild mallard duck

(*b*) Muscovy ducks

(*c*) Both Muscovy and wild mallard ducks

(*d*) None of the above

Q. 11. Immature ducks up to the age of 8-11 weeks of age are called as

(*a*) Green ducks (*b*) White ducks

(*c*) Blue ducks (*d*) Red ducks

Q. 12. Both toms and hens can attain a body weight of 1 lb by consuming

(*a*) 1.5 lbs of feed

(*b*) 2.5 lbs of feed

(*c*) 3.5 lbs of feed

(*d*) 4.5 lbs of feed

Q. 13. In rearing Turkey poults, tow clipping, de-snooding and debeaking should be done at what age

(*a*) Day old stage (*b*) 4 weeks of age

(*c*) At 6 weeks of age (*d*) At 12 weeks of age

Q. 14. Incubation period in Guinea fowl is_______days

(*a*) 25–28 days (*b*) 18–20 days

(*c*) 19–21 days (*d*) 30–35 days

Q. 15. Incubation period in Muscovy ducks is_______days

(*a*) 18 days (*b*) 21 days

(*c*) 28 days (*d*) 35 days

Q. 16. Incubation period in domestic ducks is_______days

(*a*) 18 days (*b*) 21 days

(*c*) 28 days (*d*) 35 days

Q. 17. Incubation period in Turkeys is_______days

(*a*) 18 days (*b*) 21 days

(*c*) 28 days (*d*) 35 days

Q. 18. Incubation period in Japanese quail is_______days

(*a*) 18 days (*b*) 21 days

(*c*) 28 days (*d*) 35 days

Q. 19. Incubation period in Ostrich is_______days

(*a*) 18 days (*b*) 21 days

(*c*) 28 days (*d*) 42 days

Q. 20. Incubation period in Emus is_______days

(*a*) 25 days (*b*) 30 days

(*c*) 28 days (*d*) 52 days

Q. 21. Incubation period in Pheasants is_______days

(*a*) 18-20 days (*b*) 21-22 days

(*c*) 23-25 days (*d*) 23-28 days

Q. 22. Incubation period in Pigeon is_______days

(*a*) 17-19 days (*b*) 21 days

(*c*) 28 days (*d*) None of the above

Q. 23. Incubation period in Gees is_______days

(*a*) 21-25 days (*b*) 25-28 days

(*c*) 28-34 days (*d*) None of the above

Q. 24. Incubation period in Bobwhite quail is_______days

(*a*) 23–24 days (*b*) 36-40 days

(*c*) 30-34 (*d*) None of the above

Q. 25. Chromosome number in Rhea is

(*a*) 78 (*b*) 80

(*c*) 82 (*d*) 86

II. Fill in the Blanks

1. The family names of the following species of poultry are_______, _______, _______, and_______for Guinea fowls, Turkeys, Ducks and Pea fowls, respectively.
2. A male duck is called_______
3. The duck's beak is called_______
4. Seven standard varieties of turkeys are_______, _______, _______, _______, _______, _______and_______.
5. Nine breeds of Geese are_______, _______, _______, _______, _______, _______, _______, _______ and _______.
6. Duck is the best source of Vitamin_______.
7. Turkey is generally_______in fat and calories while duck and goose are_______than chicken.
8. The ornamental type of duck breed is_______.
9. The scientific name of Japanese quail is_______.

10. The scientific name of Ducks is_______.
11. The scientific name of Muscovy Duck is_______.
12. The scientific name Ostrich is_______.
13. The scientific name of Turkey is_______.
14. The scientific name of Guinea fowl is_______.
15. The scientific name of Bobwhite quail is_______.
16. The scientific name of Goose is_______.
17. The scientific name of Emu is_______.
18. The scientific name of Rhea is_______.
19. The scientific name of Gassowary is_______.
20. Mule ducks which are sterile could be produced by crossing_______with other breeds of ducks

III. State if Statements are True or False

1. The Pekin, Runner and Campbell breeds of ducks show excessive broodiness.
2. Turkey posses 40 pairs of Autosomes.
3. The Indian Runner and Khaki Campbell are egg type breeds of ducks.
4. Duck plague is also known as duck virus enteritis
5. Duck viral hepatitis affects young ducklings (2 days to 3 weeks).
6. India has a large guinea fowl population which ranks third after chicken & ducks.
7. The principal varieties of guinea fowls are pearl, white, and lavender.

8. The average body weight of adult guinea fowl is 1 to 1.250 Kg.
9. The duck comes into production at the age of about 20–22 weeks.
10. White pekin is one of the outstanding duck breed for meat purposes.
11. The brooding period for khaki Campbell duck is 1 to 2 weeks.
12. The duck should have free access to feed without water.
13. The Khaki Campbell duck breeding was started first at Hessarghatta (Bangalore north).
14. According to American standard of perfection there are seven standard breeds of turkey.
15. The Broad Breasted Large White (BBLW) is a non-standard variety of turkey.
16. The Broad Breasted Bronze turkey originated in England.
17. The female quails are heavier (150 to 180 g) than male quails (120-150 g).
18. Duck lay more than 95% eggs before 9 am.
19. Eggs of ducks are 15 to 20 g larger than those of chicken.
20. For optimum fertility under domestication the male to female ratio should be 1:4 to 1:5 in Guinea fowl
21. The homozygotes for crest in ducks survive better.
22. In Muscovy ducks the males average nearly twice the size of the females.
23. 'Geese' is a common word which refers to both the sexes whereas 'Goose' refers to female birds and 'Gander' refers to male birds.

Answers

I. Multiple Choice Questions

1	(*a*)	8	(*b*)	15	(*d*)	22	(*a*)
2	(*c*)	9	(*c*)	16	(*d*)	23	(*c*)
3	(*b*)	10	(*a*)	17	(*c*)	24	(*a*)
4	(*b*)	11	(*a*)	18	(*c*)	25	(*b*)
5	(*a*)	12	(*c*)	19	(*d*)		
6	(*a*)	13	(*a*)	20	(*d*)		
7	(*a*)	14	(*b*)	21	(*d*)		

II. Fill in the Blanks

1. Numididae, Meleagrididae, Anatidae and Phasianidae
2. Drake
3. Bill
4. White Holland, Bourban Red, Narrgansett, Black, Slate, Bronze and Beltsville Small White
5. Toulouse, Emden, Chinese, Egyptian, Canada, Buff, Sebestopol, Pilgrim and African
6. Niacin
7. Less, Higher
8. Crested White
9. *Coturnix japonica*
10. *Anas platyrhynchos*

11. *Cairna muschata*
12. *Struthis camelus*
13. *Meleagris gallopavo*
14. *Numida meleagridis.*
15. *Colinus virginianus*
16. *Anser anser*
17. *Dromiceius novaehollandiae*
18. *Rhea americana*
19. *Casuarius casuarius*
20. Muscovy

III. True or False

1	False	7	True	13	True	19	True
2	True	8	True	14	False	20	True
3	True	9	True	15	True	21	False
4	True	10	True	16	False	22	True
5	True	11	False	17	True	23	True
6	False	12	False	18	True		

Chapter 17

Test Question Paper

I. Multiple Choice Questions (Tick the Correct Option/s)

Q. 1. Usually break in flock immunity occurs under conditions of

(*a*) Average nutritional status of the flock

(*b*) High ambient temperature

(*c*) Administration of the booster dose of vaccine

(*d*) Worm load

Q. 2. Gumbaro infection causes suppression of immunity development by destroying the function of

(*a*) T-cells

(*b*) Suppression in the production of immunoglobulin

(*c*) B-cells

(*d*) Interferes with the work of lymphocytes

Q. 3. One of the following disease causes production of rough shelled eggs

(*a*) Ranikhet (*b*) CDS-76

(*c*) Infectious laryngotrachitis (*d*) Infectious bronchitis

Q. 4. In one of the virus disease infection the mild vaccine gives protection to ailing birds

(*a*) Infectious bursal disease (*b*) Marek's disease
(*c*) New Castle disease (*d*) CDS-76

Q. 5. The environmental temperature beyond 40 °C exhibits adverse effects on production traits via

(*a*) Low feed consumption
(*b*) Alkalosis
(*c*) Increase in cloacal temperature of hen
(*d*) Low blood level of calcium
(*e*) All of the above

Q. 6. The lethal aflatoxicity in feed is developed at minimum humidity level of

(*a*) 3% (*b*) 13%
(*c*) 17% (*d*) 26%
(*e*) None of the above

Q. 7. The birds having following type of social pecks order lays more eggs

(*a*) Dominant (*b*) Ressive
(*c*) Middle peck order (*d*) None of the above

Q. 8. Sundried poultry excreta contain percentage protein

(*a*) 13 (*b*) 29
(*c*) 35 (*d*) 24

Q. 9. One hen can on an average digest percentage of feed eaten within 24 hours

(*a*) 90 percent and above (*b*) 80 percent and above
(*c*) 70 percent and above (*d*) None of the above

Q. 10. If the liver of chicken on post mortem appears blackish in colour with rounded borders, the disease most probably suspected is

(*a*) Marek's disease (*b*) Salmonellosis
(*c*) Colibacillosis (*d*) Big liver disease

Q. 11. The layer pullet should be given pre-layer ration form the following age if starts its production in the 20th week of age.

(*a*) On completion of 20th week

(*b*) On completion of the 18th week

(*c*) On completion of the 16th week

(*d*) Not necessary to give pre-layer ration

Q. 12. In summer stress the requirement of following vitamin increases

(*a*) Vitamin E

(*b*) Vitamin B_6

(*c*) Vitamin K_3

(*d*) All of the above

Q. 13. Under natural conditions the development of embryo starts at environmental temperature of

(*a*) 13.5°C (*b*) 21.5°C

(*c*) 27.5°C (*d*) None of the above

Q. 14. The liquefaction of the albumen on storage of egg is caused by

(*a*) pH of the egg goes beyond eight

(*b*) Escape of CO_2 from egg

(*c*) Change in osmotic pressure

(*d*) (*a*) and (*b*)

(*e*) (*a*), (*b*) and (*c*)

Q. 15. The shape index value normal shelled eggs is in the range of

(*a*) 90-100 (*b*) 80-90

(*c*) 70-80 (*d*) 60-70

Q. 16. The weight of the leg muscle is close to percentage of the dressed carcass weight

(*a*) 22 (*b*) 26

(*c*) 28 (*d*) 32

Q. 17. During majority of the incubation period the respiratory gaseous exchange takes place via

(*a*) Shell membrane

(*b*) Amnion

(*c*) Yolk–sac membrane

(*d*) Chorio-allantoic membrane

(*e*) All of the above

Q. 18. Maintenance energy requirement by laying hen per kg body weight is approximately

(*a*) 100 $W^{0.75}$ kcal

(*b*) 120 $W^{0.75}$ kcal

(*c*) 135 $W^{0.75}$ kcal

(*d*) 80 $W^{0.75}$ kcal

Q. 19. Broiler breeders can be raised economically observing quantitative feed restriction of the percentage *ad libitum* feeding without affecting growth and production

(*a*) 10% (*b*) 20%

(*c*) 25% (*d*) 30%

Q. 20. Feather colour is affected by the deficiency of amino acid in diet

(*a*) Arginine (*b*) Lysine

(*c*) Tryptophane (*d*) Methionine

Q. 21. Vitamin A is stable when mixed with

(*a*) Vitamin and mineral premix

(*b*) Vitamin and Ground nut cake

(*c*) Vitamin with limestone powder

(*d*) None of the above

Q. 22. Vitamin C is the member of

(*a*) Electron transport system (*b*) ATP

(*c*) Acetylcholine (*d*) Acetyl-Co-A

Q. 23. The drift variance in the control population increases with

(*a*) Generation number

(*b*) Error in the measurement of genetic variance

(*c*) Genotype–environment interaction

(*d*) Change in the variance of the population

Q. 24. Genotype and environmental interaction in restricted sense would decrease selection response if

(*a*) Heritability of the trait is low

(*b*) Genetic correlation between trait is negative

(*c*) Repeatability of the trait is negative

(*d*) None of the above

Q. 25. Following gene is situated on the autosomal chromosome of karyotype of chicken

(*a*) Silver

(*b*) Naked neck

(*c*) Inhibition of dermal pigment

(*d*) None of the above

Q. 26. The fertility of the leghorn population is partially influenced under homologous state of

(*a*) Frizzle gene

(*b*) Rose comb

(*c*) Shaker gene

(*d*) Sex-linked silver gene

Q. 27. The diallel crossing is one of the following

(*a*) Mating procedure

(*b*) Breeding aid to improve livestock

(*c*) Selection aid

(*d*) Comparative crossbred combination evaluation aid

Q. 28. The high frequency mating by the male is the sign of

(*a*) High fertilization capacity

(*b*) More semen deposition in oviduct

(*c*) Dominance of male

(*d*) All the above

Q. 29. The main source of energy the developing embryo derives during last phase of development is from

(*a*) Yolk sac (*b*) Carbohydrate

(*c*) Fat (*d*) All of the above

Q. 30. Maize is deficient in the following amino acids

(*a*) Methionine and Lysine

(*b*) Methionine and Argine

(*c*) Lysine and arginine

(*d*) Lysine and Tryptophane

Q. 31. Dietary Vitamin A causes the 1st peak of embryonic mortality due to the failure of

(*a*) Oxygen supply

(*b*) Cellular multiplication

(*c*) To develop capillary island

(*d*) Mobilize fat

Q. 32. In random mating population the chance of the following is minimized

(*a*) Differential hatchability

(*b*) Differential selection intensity

(*c*) Differential fertility

(*d*) Differential genetic correlation

Q. 33. In idealistic population the family selection cannot be better than the individual selection if

(*a*) Heritability is low

(*b*) Selection intensity is high

(*c*) Heritability of family is larger than the individual basis

(*d*) None of the above

Q. 34. Recurrent selection scheme to utilize the non-additive genetic variance was proposed by

(*a*) Dickerson (1965) (*b*) Griffin (1956)

(*c*) Sparge and Tatum (1956) (*d*) Crow (1953)

Q. 35. Ranikhet vaccination can be done by the following route

(*a*) Intra muscular (*b*) Intra peritoneal

(*c*) Subcutaneous (*d*) All of the above

Q. 36. The mode of transmission of infectious Coryza is

(*a*) Horizontal (*b*) Vertical

(*c*) Manageraial (*d*) None of the above

Q. 37. Less availability of energy from wheat is due to

(*a*) Cellulose covering (*b*) Pentosans

(*c*) Glycosides (*d*) Trypsin inhibitor

Q. 38. Excess dietary calcium may facilitate the availability of

(*a*) Manganese (*b*) Vitamin A

(*c*) Choline chloride (*d*) None of the above

Q. 39. The activity to pip the shell is initiated by the change in the supply within the egg

(*a*) Water (*b*) Oxygen

(*c*) Carbon dioxide (*d*) All of the above

Q. 40. The Gumbaro disease may cause mortality in birds of the following age group:

(*a*) 3-6 weeks of age

(*b*) 12-16 weeks of age

(*c*) Any age group from 3 to 16 weeks of age

(*d*) None of the above

II. Fill the Blanks

1. In the presence of vitamin D_3 the parathyroid hormone assists to maintain the_______level of blood.

2. Vitamin A maintains the_______of the lipoprotein membrane of the skin.

3. The mineral element does not require_______for the digestion in gut.

4. The sunflower cake is good protein source but contains high level of_______

5. Recently the Central avian Research Institute, Izatnagar has released commercial broiler named_______for farmers use.

6. The colour of Kadaknath breed is bluish black due to presence of_______gene.

7. The cell in which the fat is deposited as triglycerides is known as_______

8. The inner lining of the gizzard is formed by the hardening of_______secretion.

9. The mineral content of shelled egg on dry matter basis is on an average_______percent.

10. The major portion of the shelled egg glycoprotein is contributed by_______fraction of albumin.

11. Movement of the embryo by turning in the incubator in the first week of age is important for the structuring of_______

12. Shell pores of the egg changes their size during summer heat due to the failure in the formation of_______

13. In mycoplasma infection in hatcheries the hatching eggs are commonly dipped for_______minutes in solution containing_______

14. Plastic covering on hatching egg containers flushed with inert gas is not used for_______duration of storage.

15. Parrot beak is developed in embryo due to the deficiency of_______in maternal diet.

16. Naked neck gene is situated on chromosome No._______of the chicken karyotype.

17. The feed efficiency to grow kg body weight during 0-7 days of age is_______as compared with that for 14-21 days period.

18. The pullet flock that start laying at the age of 16 weeks would produce_______number of eggs than those starting production at 21 weeks of age.

19. Developing embryo is choosy to utilize_______fatty acid more efficiently for its developmental energy requirement for growth.

20. Chick through ornithine cycle can synthesize ornithine but cannot synthesize Choline due to lack of enzyme_______

21. Choline interferes with the synthesis of cystine from_______

22. Lysine amino acid is destroyed by radiation in compounded feed due to the formation of_______complex.

23. On an average the amino acids present in the feed are available_______percent to chick.

24. Menadione sodium bisulphate along with antibiotics and vitamins are used to improve_______of birds.

25. The enzyme in whose presence the two carbon structure of amino acids join by the process of transamination is_______

26. The electro-chemical potential difference (ECPD) regulates the absorption of_______from intestine in to blood.

27. Chelating agents enhance the absorption of_______from intestine.

28. Black colour of the sunflower cake is due to the presence of_______acid.

29. The inbuilt weakness of selection index theory is the assumption of_______

30. The main traits on which maximum emphasis is given in selection of male broiler parents are_______and_______

III. Mark the Statement as True (T) or False (F)

1. Linoleic acid is the major fatty acid found in the shelled egg.
2. Poultry excreta which is sun dried can easily be recycled at the rate of 5-10 percent in diet of growers.
3. The percentage of defective eggs is low in broiler breeder parents in early phase of production.
4. Under appropriate humidity of the Hatcher the chick pip the egg shell below the air cell membrane.
5. Gumboro disease under normal bio-security arrangements may cause 10-15% loss of growing pullets beyond the age of 16 weeks.
6. The hatching eggs in incubator can safely be fumigated by 1 x concentration of formalin and potassium permanganate at the age between 12 to 48 hours post setting.
7. Get away system of cage housing is more inhumane system of housing than of colony cage system.
8. The colour of feather is mainly due to the presence of melanin.
9. As the incubator setter size will increase the temperature of setter will also to be increased to obtain good results.
10. Tannins are mostly found in millet grains.
11. Mycoplasmosis has transovarian transmission.
12. The wing and the leg paralysis in chicks aged 2-5 day age is seen in viral encephalomyelitis.
13. The commercial layer pullets usually show visceral form of Marek's disease between 18-22 weeks of age.
14. The Naked neck gene cause low fertility in homozygous state.

15. The main use of sex-linked dwarfing gene is to develop male broiler line for the commercial chick production.

16. Intermediate IBD vaccines are not as useful as that the lukert strain in the preventive vaccination against Gumboro disease.

17. The pH of cock semen is more towards alkaline side.

18. In majority of the incubation period the gaseous exchange takes place via allantoic membrane.

19. Eumelanin pigment gives black and blue pigmentation to feathers.

20. High laying intensity could be obtained by raising pullets under ahemral lighting schedule.

21. Theoretically the heterosis will be expected to be zero when the trait is controlled by the additive genetic variance alone.

22. Hen day egg production is better indicator of flocks economic viability than the hen housed egg production.

23. Retro viruses are commonly used as transmitting agent of DNA for the development of transgenic animals.

24. The desirable pH of drinking water for the poultry should be towards the acidic side.

25. The maximum concentration of ammonia in the shed should not increase beyond 30 ppm.